Conference Proceedings of the Society for Experimental Mechanics Series

Series Editor

Kristin B. Zimmerman, Ph.D.
Society for Experimental Mechanics, Inc.,
Bethel, CT, USA

The Conference Proceedings of the Society for Experimental Mechanics Series presents early findings and case studies from a wide range of fundamental and applied work across the broad range of fields that comprise Experimental Mechanics. Series volumes follow the principle tracks or focus topics featured in each of the Society's two annual conferences: IMAC, A Conference and Exposition on Structural Dynamics, and the Society's Annual Conference & Exposition and will address critical areas of interest to researchers and design engineers working in all areas of Structural Dynamics, Solid Mechanics and Materials Research.

More information about this series at http://www.springer.com/series/8922

Martha E. Grady
Editor

Mechanics of Biological Systems and Materials & Micro-and Nanomechanics, Volume 4

Proceedings of the 2019 Annual Conference on Experimental and Applied Mechanics

Springer

Editor
Martha E. Grady
Department of Mechanical Engineering
University of Kentucky
Lexington, KY, USA

ISSN 2191-5644 ISSN 2191-5652 (electronic)
Conference Proceedings of the Society for Experimental Mechanics Series
ISBN 978-3-030-30012-8 ISBN 978-3-030-30013-5 (eBook)
https://doi.org/10.1007/978-3-030-30013-5

This Springer imprint is published by the registered company Springer Nature Switzerland AG.
The registered company address is: Gewerbestrasse 11, 6330 Cham, Switzerland

Preface

Mechanics of Biological Systems and Materials & Micro- and Nanomechanics represents one of six volumes of technical papers presented at the 2019 SEM Annual Conference and Exposition on Experimental and Applied Mechanics organized by the Society for Experimental Mechanics and held in Reno, NV, June 3–6, 2019. The complete Proceedings also include volumes on *Dynamic Behavior of Materials; Challenges in Mechanics of Time-Dependent Materials, Fracture, Fatigue, Failure and Damage Evolution; Advancement of Optical Methods & Digital Image Correlation in Experimental Mechanics; Mechanics of Composite, Hybrid and Multifunctional Materials; and Residual Stress, Thermomechanics & Infrared Imaging and Inverse Problems.*

Each collection presents early findings from experimental and computational investigations on an important area within Experimental Mechanics, the Mechanics of Biological Systems and Materials, Micro- and Nanomechanics, and other experimental and applied mechanics such as research in progress.

The Biological Systems and Materials segment of this volume summarizes the exchange of ideas and information among scientists and engineers involved in the research and analysis of how mechanical loads interact with the structure, properties, and function of living organisms and their tissues. The scope includes experimental, imaging, numerical, and mathematical techniques and tools spanning various length and time scales. Establishing this symposium at the Annual Meeting of the Society for Experimental Mechanics provides a venue where state-of-the-art experimental methods can be leveraged in the study of biological and bio-inspired materials, traumatic brain injury, cell mechanics, and biomechanics in general. A major goal of the symposium was for participants to collaborate in the asking of fundamental questions and the development of new techniques to address bio-inspired problems in society, human health, and the natural world. The 2019 Symposium is the 9th International Symposium on the Mechanics of Biological Systems and Materials. The organizers would like to thank all the speakers and staff at SEM for enabling a successful program.

The Micro- and Nanomechanics segment of this volume focuses on specialized scientific areas that involve miniaturizing conventional scale components and systems to take advantage of reduced size and weight and/or enhanced performance or novel functionality. These fields also encompass the application of principles ranging from the micron scale down to individual atoms. Sometimes these principles borrow from conventional scale laws but often involve new physical and/or chemical phenomena that require new behavioral laws and impart new properties to exploit. Studying how mechanical loads interact with components of these scales is important in developing new applications, as well as assessing their reliability and functionality. Establishing this symposium at the Annual Meeting of the Society for Experimental Mechanics provides a venue where state-of-the-art experimental methods can be leveraged in these endeavors.

The 2019 Symposium is the 20th in the series and addresses pertinent issues relating to design, analysis, fabrication, testing, optimization, and applications of micro- and nanomechanics, especially as these issues relate to experimental mechanics of microscale and nanoscale structures.

It is with deep gratitude that we thank the Organizing Committee, Session Chairs, Authors and Keynote Speakers, Participants, and SEM Staff for making the 20th *International Symposium on Micro- and Nanomechanics (ISMAN) and the 9th International Symposium on the Mechanics of Biological Systems and Materials* a valuable and unforgettable experience.

University of Kentucky, USA — Martha Grady
University of Wisconsin-Madison, USA — Jacob Notbohm
University of Wisconsin-Madison, USA — Christian Franck
Air Force Research Laboratory, USA — LaVern Starman
Nanomechanics Inc., USA — Jenny Hay

Contents

Terahertz Protein Vibrations: The Usefulness of Coarse-Grained Numerical Models

Giuseppe Lacidogna, Domenico Scaramozzino, Gianfranco Piana, and Alberto Carpinteri

Abstract Understanding the way in which proteins vibrate in their folded state is pivotal for a broad comprehension of their biological activity. In particular, vibrations in the terahertz range are indicated in the current literature as being involved in protein conformational changes. Nowadays, frequencies around or below 1 THz can be detected for example by Raman spectroscopy using proper ultra-low frequency filters. In previous studies, some of the authors performed modal analysis of all-atom lattice models to investigate the expansion-contraction mode shapes associated to low-frequency Raman peaks detected experimentally on lysozyme and Na^+/K^+-ATPase powder samples. In this contribution, all-atom calculations are compared to new ones derived from a simplified coarse-grained mechanical model; the latter was built-up considering only C_α atoms, i.e., the protein backbone. The efficacy of the coarse-grained model in describing delocalized and global expansion-contraction protein vibrations as well as its limitations are discussed.

Keywords Protein vibrations · Modal analysis · Coarse-grained model · Raman spectroscopy · THz range

Introduction

Protein three-dimensional structure is known to be strictly related to biological functionality, which is performed in a dynamic fashion. Therefore, analyzing the way in which proteins vibrate around the native state is of utmost importance, in order to comprehend the complex mechanisms hidden behind protein activity. Vibrational motions can occur at different scales, e.g., involving amino acid side chains, peptide bonds, secondary structures, as well as the whole protein. They are completely defined by the corresponding displacement field and frequency of vibration. Generally, the smaller the scale, the higher the frequency. Among all the vibrational modes, the low-frequency ones have received increasing attention by the scientific community, as they have been indicated to be involved in protein conformational changes [1]. These motions affect large portions of the protein and are believed to play a crucial role in controlling binding activity, which is expressed through the opening and closing of specific clefts.

Protein dynamics can be effectively investigated by means of both numerical and experimental tools. As for the former, molecular dynamics simulations and normal mode analysis [2–5] are generally employed, by using all-atom and coarse-grained models. Each modeling procedure has its own advantages and drawbacks, and leads to different levels of approximation. Anyway, it has been shown that even simplified models, such as coarse-grained ones based on native topology, are able to capture the essential protein behavior [6, 7].

From an experimental viewpoint, protein vibrations can be detected by means of spectroscopy techniques, such as THz-TDS (terahertz time-domain spectroscopy) [8] and Raman spectroscopy. In particular, the latter has proven to be a powerful tool in detecting vibrations around or below 1 THz (~30 cm^{-1}), by using specific ULF (ultra-low frequency) filters. In previous studies, some of the authors made use of ULF-Raman spectroscopy on lysozyme [9] and Na^+/K^+-ATPase [10] powder samples, obtaining some peaks around 0.8 THz. Modal analysis (i.e., linear normal mode calculations) was performed to investigate the expansion-contraction dynamics of all-atom lattice models [9, 11]. The mode shapes associated to the Raman peaks were found to involve the whole lysozyme structure and a large portion of Na^+/K^+-ATPase.

In this contribution, a simplified coarse-grained mechanical model is proposed, which is based only on C_α atoms, aimed at investigating the low-frequency expansion-contraction protein vibrations. The effectiveness of the model is shown by comparing the results with those deriving from previous all-atom simulations, both regarding vibrational frequencies and mode shapes. Its limitations are discussed as well.

G. Lacidogna · D. Scaramozzino (✉) · G. Piana · A. Carpinteri
Politecnico di Torino, Department of Structural, Geotechnical and Building Engineering, Torino, Italy
e-mail: domenico.scaramozzino@polito.it

M. E. Grady (ed.), *Mechanics of Biological Systems and Materials & Micro-and Nanomechanics, Volume 4*,
Conference Proceedings of the Society for Experimental Mechanics Series, https://doi.org/10.1007/978-3-030-30013-5_1

Coarse-Grained Mechanical Model and Modal Analysis

The proposed coarse-grained mechanical model aims at focusing on low-frequency expansion-contraction modes around protein native state. As mentioned in the Introduction, these motions generally involve large portions of the protein, therefore it seems unnecessary to take into account local details such as, for example, amino acid side chains. For this reason, based on PDB (Protein Data Bank) information [12], only C_α atoms positions have been used for the construction of the model.

In order to analyze the dynamic behavior of the protein backbone, a cut-off distance of 4 Å has been set (i.e., C_α atoms are connected only if their distance is lower than the imposed cut-off value), since generally the distance between two consecutive C_α atoms is 3.8 Å. The protein structure turns out to be a lattice model, made up of nodes, corresponding to C_α atoms, and links, corresponding to C_α–C_α bonds. Two main parameters are still needed to completely define the mechanical model: node mass and interatomic bond stiffness. As for the former, the total mass of the protein has been equally divided among C_α atoms; as for the latter, it has been derived from the rigidity of covalent bonds of the protein backbone (–C_α–C–N–C_α–).

As shown in Fig. 1, each C_α–C_α link must in a sense simulate the effect of three covalent bonds. Considering for C–C and C–N bonds the stiffness and equilibrium length reported in Table 1, one can compute the mean values, which are supposed to belong to an "ideal" backbone bond.

Assuming that the stiffness of an "ideal" bond is inversely proportional to the bond length, one can finally compute the stiffness of the C_α–C_α link through Eq. (1):

$$k_{C_\alpha - C_\alpha} = k_{backbone} \frac{L_{backbone}}{L_{C_\alpha - C_\alpha}}, \quad (1)$$

where $k_{backbone}$ and $L_{backbone}$ are reported in Table 1 and $L_{C\alpha-C\alpha}$ is computed depending on the actual positions of the C_α atoms. In accordance with [9, 11], Eq. (1) provides the axial stiffness of interatomic bonds, modeled as elastic bars. This stiffness value is different for different links, based on their length, contrarily from single-parameter approaches adopted in other works [2–4]. Built the coarse-grained FE (Finite Element) model and assigned the mechanical properties, Lusas software [13] is used to perform modal analysis. It basically consists in solving the following equation:

$$\left(\boldsymbol{K} - \omega_n^2 \boldsymbol{M}\right) \cdot \boldsymbol{\delta}_n = \boldsymbol{0}, \quad (2)$$

where $\boldsymbol{K}$ and $\boldsymbol{M}$ are the global stiffness and mass matrices, respectively; $\omega_n{}^2$ represents the nth eigenvalue, which is related to the nth frequency of vibration f_n by the following formula:

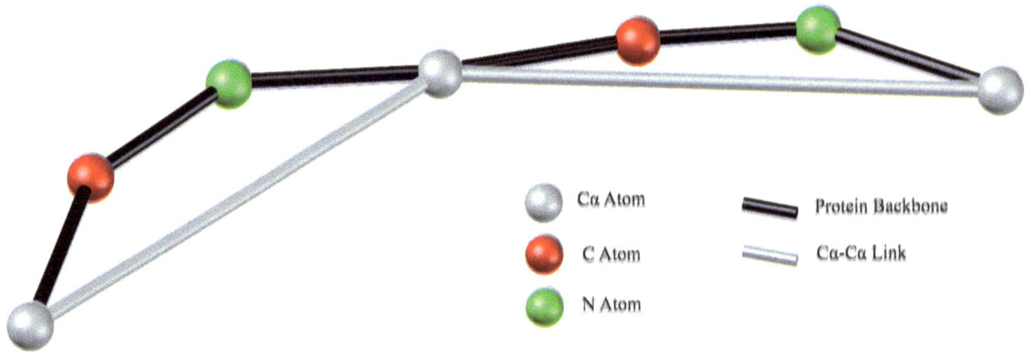

Fig. 1 Scheme for the derivation of C_α–C_α bond stiffness

Table 1 Covalent bond parameters

Covalent bond	Stiffness [N/m]	Equilibrium length [Å]
C–C	180	1.54
C–N	160	1.47
Backbone—Mean[a]	166.7	1.49

[a]Considering 1 C–C and 2 C–N bonds between two consecutive C_α atoms

$$f_n = \frac{\omega_n}{2\pi}, \tag{3}$$

and $\boldsymbol{\delta}_n$ is the corresponding eigenvector, i.e., the displacement field of the nth vibrational mode. Note that, since protein structure is not externally constrained, the first six vibrational modes are rigid motions with zero-frequency. In the following section, the results deriving from modal analysis are shown, with respect to lysozyme and Na^+/K^+-ATPase.

Results and Discussion

In previous works [9–11] some of the authors performed Raman spectroscopy measurements on lysozyme and Na^+/K^+-ATPase samples, with special ULF filters, and detected strong peaks around 0.8 THz. The obtained spectra, with focus on the low-frequency region, are shown in Fig. 2. By means of modal analysis applied to all-atom models, it was found that in the frequency range near 0.8 THz lysozyme exhibits very collective motions [9], whereas some delocalized vibrations, involving the protein ends, are identified as regards Na^+/K^+-ATPase [11].

Here, we performed modal analysis by means of the mechanical coarse-grained model proposed in the previous section, and compared the results with those arising from all-atom calculations. In Fig. 3, the all-atom and coarse-grained lattice structures are reported for both lysozyme (pdb code: 4ym8) and Na^+/K^+-ATPase (2zxe). It is noteworthy that, by using the coarse-grained model, there was approximately an eight-fold reduction in the number of nodes with respect to the all-atom representation, thus leading to a remarkable saving in terms of computational cost: the reduction lies from 1000 to 129 nodes for lysozyme, and from 10,133 to 1296 for Na^+/K^+-ATPase. Moreover, an easier interpretation of the results in terms of vibration modes was achieved at the same time.

By applying modal analysis on both all-atom and coarse-grained models, we obtained all the vibrational frequencies and the mode shapes. In Fig. 4, the comparison between the results is shown in terms of eigenfrequencies (expressed in THz). As can be seen, in the range 0.1–4.5 THz, the obtained eigenfrequencies almost coincide, i.e., on average both models provided approximately the same frequency value for each eigenmode. Therefore, in order to investigate the vibrational modes lying in a frequency range which can be associated to an experimental Raman peak (Fig. 1), one can use more conveniently the coarse-grained model, if the peak is found up to 4.5 THz (~150 cm^{-1}). Contrariwise, for higher frequencies (above 5 THz) the correlation between the all-atom and coarse-grained results gets lost since local motions dominate, such as vibration of amino acid side chains, which in turn cannot be captured by the coarse-grained model.

It is interesting to observe also the comparison in terms of modal displacements. This can be achieved by means of Modal Assurance Criterion (MAC) [14], which is defined according to the following equation:

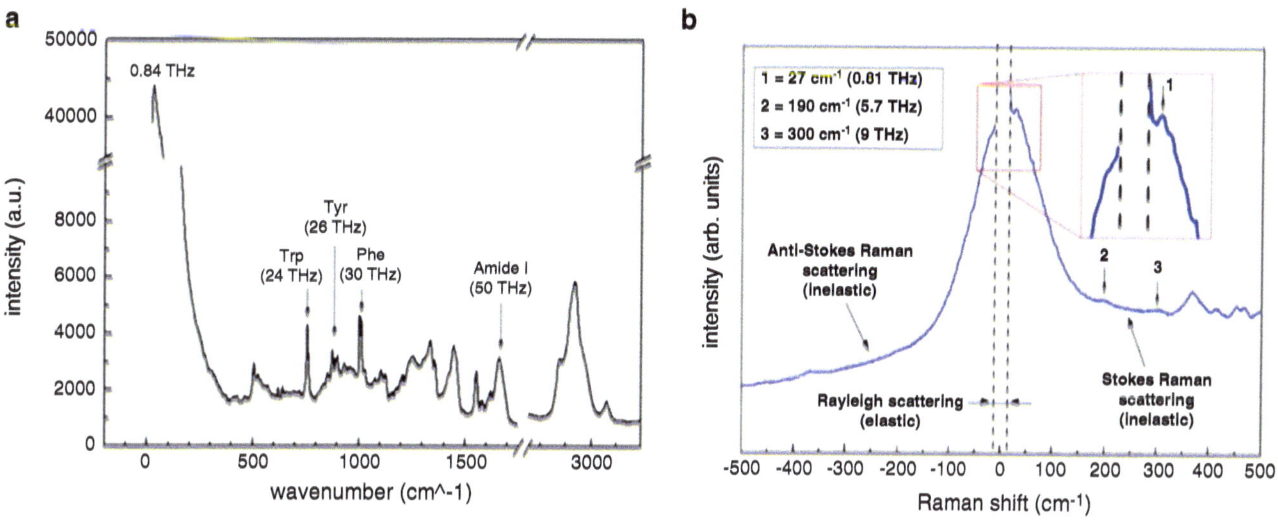

Fig. 2 Raman spectra of (**a**) lysozyme [9] and (**b**) Na^+/K^+-ATPase [11]

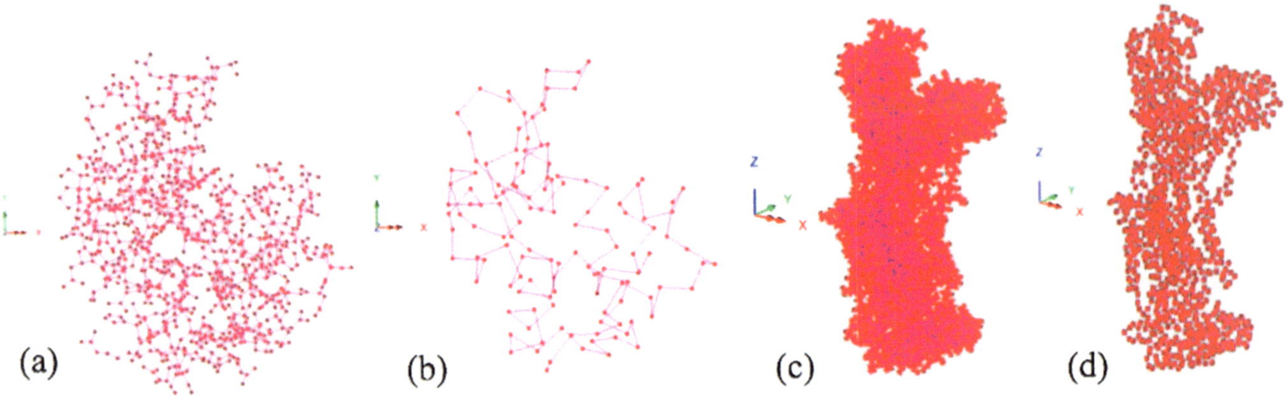

Fig. 3 All-atom model of (**a**) lysozyme and (**c**) Na^+/K^+-ATPase; coarse-grained model of (**b**) lysozyme and (**d**) Na^+/K^+-ATPase

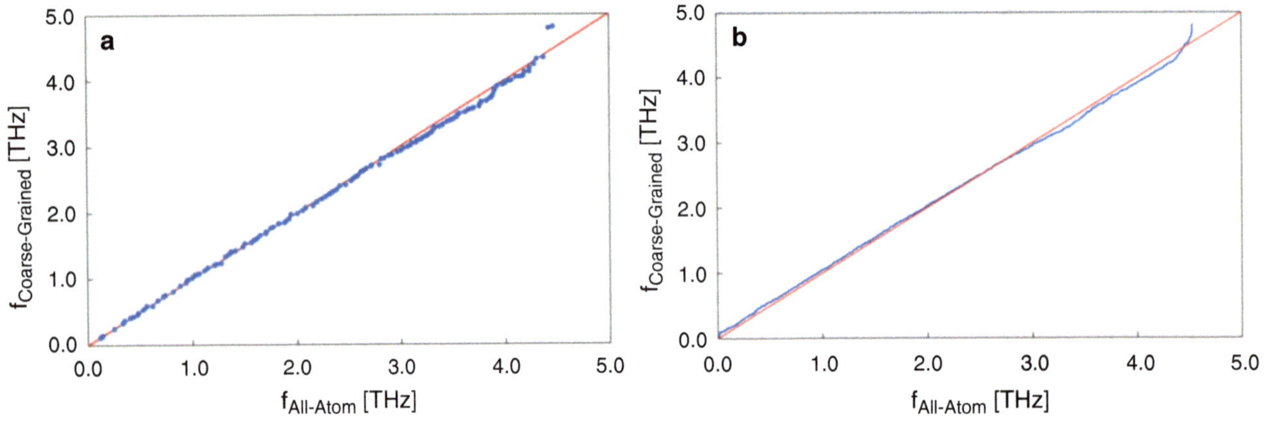

Fig. 4 Frequency comparison between coarse-grained and all-atom model for (**a**) lysozyme and (**b**) Na^+/K^+-ATPase

$$MAC_{i,j} = \frac{\left(\boldsymbol{\delta}_{i,CG}{}^{T}\boldsymbol{\delta}_{j,AA}\right)^{2}}{\left(\boldsymbol{\delta}_{i,CG}{}^{T}\boldsymbol{\delta}_{i,CG}\right)\left(\boldsymbol{\delta}_{j,AA}{}^{T}\boldsymbol{\delta}_{j,AA}\right)}, \tag{4}$$

where $\boldsymbol{\delta}_{i,CG}$ and $\boldsymbol{\delta}_{j,AA}$ represent the ith and j^{th} modal shape obtained by coarse-grained and all-atom calculations, respectively. Obviously, as far as $\boldsymbol{\delta}_{j,AA}$ is concerned, only the displacements of C_α atoms are considered within the vector. For example, regarding the seventh vibrational modes of lysozyme ($f_{7,CG} = 0.114$ THz, $f_{7,AA} = 0.118$ THz), one obtains $MAC_{7,7} = 0.985$, which is also evident from Fig. 5, which shows the normalized displacements for each residue, i.e. C_α atom.

In Fig. 6, the MAC matrix deriving from the calculation based on Eq. (4) is shown for the first 30 pairs of modes (not including the six rigid motions). Each grid of the matrix, of coordinates *i* and *j*, shows the MAC value (in color scale) between the ith coarse-grained and the jth all-atom eigenmode. As can be seen, high MAC values are obtained within the matrix diagonal, especially for the low-frequency eigenmodes, where values higher than 0.9 are generally obtained. Otherwise, for higher mode numbers, MAC values decrease, as well as they spread around the diagonal, since some interaction appears between different modes. As far as Na^+/K^+-ATPase is concerned, the results deriving from the MAC analysis based on Eq. (4) cannot be easily interpreted, since the transmembrane pump is made up of three different amino acid chains. Therefore, although the range of the obtained all-atom and coarse-grained eigenfrequencies is almost the same (Fig. 4b), the exact correspondence between the eigenmodes cannot be achieved, since different chains can vibrate alternatively at similar frequencies.

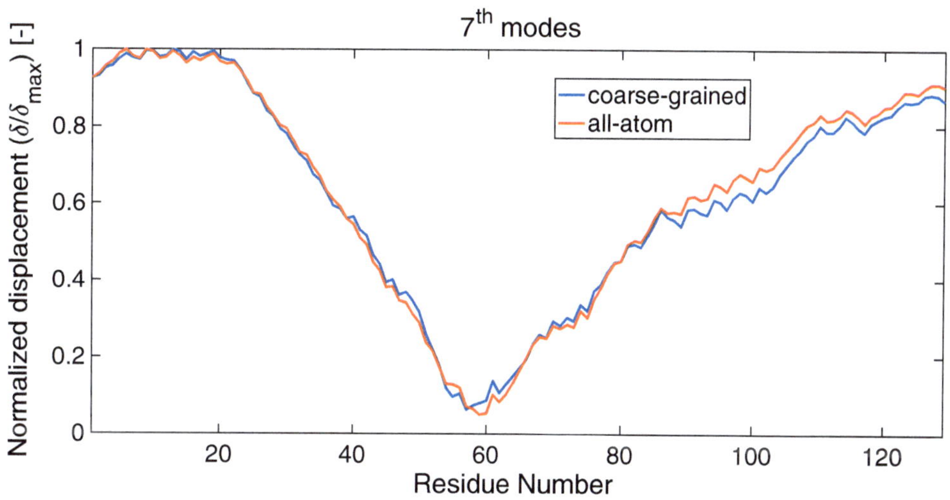

Fig. 5 Modal shapes comparison for lysozyme: seventh modes ($MAC_{7,7} = 0.985$)

Fig. 6 MAC matrix for the first 30 pairs of modes of lysozyme

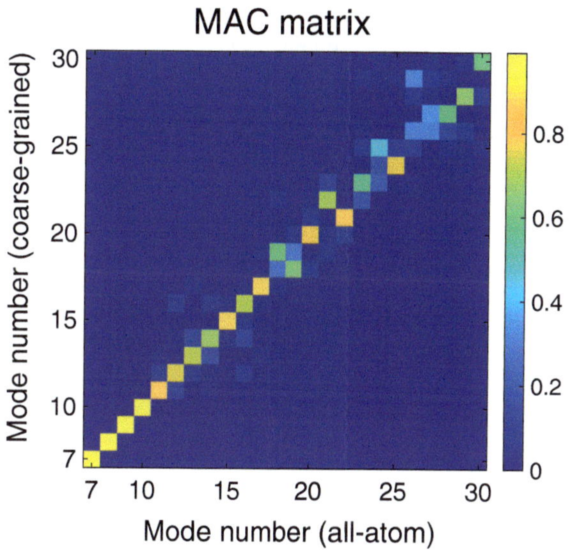

Conclusions

In this contribution, we presented a simplified coarse-grained model, based only on C_α atoms, aimed at focusing on focusing on low-frequency expansion-contraction protein dynamics. This model was developed to analyze the vibrations of protein backbone, neglecting long-range interactions. Lysozyme and Na^+/K^+-ATPase were modeled in LUSAS finite element code: the comparison with all-atom calculations confirmed that, when global or delocalized motions occur, frequencies in the terahertz range are involved and the coarse-grained model is sufficient to capture the essential expansion-contraction dynamics. In particular, the results arising from all-atom and coarse-grained calculations show an impressive agreement both in terms of eigenfrequencies and, at least for the low-frequency modes, modal shapes as far as single-chain proteins are involved, like in the case of lysozyme. However, when it comes to more complex structures, made up of several amino acid chains, such as Na^+/K^+-ATPase, the two models provide approximately the same values in terms of eigenfrequencies, but the exact correspondence between the mode shapes cannot be found.

References

1. S. Mahajan, Y.H. Sanejouand, On the relationship between low-frequency normal modes and the large-scale conformational changes of proteins. Arch. Biochem. Biophys. **567**, 59–65 (2015)
2. M.M. Tirion, Large amplitude elastic motions in proteins from a single-parameter, atomic analysis. Phys. Rev. Lett. **77**, 1905–1908 (1996)
3. I. Bahar, A.R. Atilgan, B. Erman, Direct evaluation of thermal fluctuations in proteins using a single-parameter harmonic potential. Fold. Des. **2**, 173–181 (1997)
4. A.R. Atilgan, S.R. Durell, R.L. Jernigan, M.C. Demirel, O. Keskin, I. Bahar, Anisotropy of fluctuation dynamics of proteins with an elastic network model. Biophys. J. **80**, 505–515 (2001)
5. A. Nicolai, P. Delarue, P. Senet, Low-frequency, functional, modes of proteins all-atom and coarse-grained normal mode analysis, in *Computational methods to study the structure and dynamics of biomolecules and biomolecular processes. Springer series in bio-/neuroinformatics*, ed. by A. Liwo, (Springer, Berlin, Heidelberg, 2014)
6. V. Tozzini, Coarse-grained models for proteins. Curr. Opin. Struct. Biol. **15**, 144–150 (2005)
7. I. Bahar, A.J. Rader, Coarse-grained normal mode analysis in structural biology. Curr. Opin. Struct. Biol. **15**, 586–592 (2005)
8. G. Acbas, K.A. Niessen, E.H. Snell, A.G. Markelz, Optical measurements of long-range protein vibrations. Nat. Commun. **5**, 3076 (2014)
9. A. Carpinteri, G. Lacidogna, G. Piana, A. Bassani, Terahertz mechanical vibrations in lysozyme: Raman spectroscopy vs modal analysis. J. Mol. Struct. **1139**, 222–230 (2017)
10. G. Lacidogna, G. Piana, A. Bassani, A. Carpinteri, Raman spectroscopy of Na/K-ATPase with special focus on low-frequency vibrations. Vib. Spectrosc. **92**, 298–301 (2017)
11. A. Carpinteri, G. Piana, A. Bassani, G. Lacidogna, Terahertz vibrational modes in Na/K-ATPase. J. Biomol. Struct. Dyn. **37**, 256–264 (2019)
12. Protein Data Bank. https://www.rcsb.org
13. http://www.lusas.com
14. R.J. Allemang, D.L. Brown, A correlation coefficient for modal vector analysis. *Proceedings of the 1st IMAC*, Orlando, FL, USA, 1982

Protein Conformational Changes and Low-Frequency Vibrational Modes: A Similarity Analysis

Domenico Scaramozzino, Giuseppe Lacidogna, and Alberto Carpinteri

Abstract The study of protein vibration and dynamics is receiving increasing attention among researchers, both from a numerical and experimental perspective. By using terahertz spectroscopy techniques, it has been shown that conformational changes, crucial for protein biological function, are strictly related to low-frequency vibrational modes. These motions generally occur in the terahertz range (~0.1–2 THz) involving large portions of the protein. The present contribution aims at investigating the role of terahertz (expansion-contraction) vibrational modes to protein conformational change from a numerical viewpoint. Modal analysis is performed by using C_α-only coarse-grained mechanical models: the obtained mode shapes are compared, by means of three similarity indexes, to the displacement field of protein conformational change. In particular, lysine-arginine-ornithine (LAO) binding protein is selected as a case study.

Keywords Conformational change · THz vibrational modes · Modal analysis · Similarity indexes · LAO binding protein

Introduction

Every aspect of protein biological activity is ruled by conformational changes. These structural rearrangements of protein three-dimensional shape occur in ligand binding phenomena, signaling and transportation processes, protein-protein interactions, etc. For example, hemoglobin changes its shape when binding to oxygen molecules, switching from a deoxy- to an oxy-state. Molecular motors, such as kinesin and myosin, move on microtubules and actin filaments, respectively, because of large conformational changes in their motor heads caused by ATP hydrolysis. Vinculin-talin complex at focal adhesions is driven by conformational switches in vinculin that expose cryptic binding sites [1].

However, given the strict relationship between biological functionality and protein native shape, conformational changes should have their fundamental reasons in some intrinsic feature of protein structure. In particular, low-frequency vibrational modes are believed to represent the ideal candidates for governing such transitions [2]. These motions generally involve the whole protein structure, or large portions of it. Recently, some of the authors performed modal analysis on protein mechanical models, aimed at investigating the expansion-contraction motions, by using both all-atom [3, 4] and coarse-grained representation [5], and obtained that these mechanical vibrations occur in the terahertz range.

From an experimental viewpoint, low-frequency vibrations can be detected, among others, by Raman spectroscopy technique. Brown et al. [6] analyzed α chymotrypsin samples and found that these vibrational motions involve the whole protein structure. More recently, some of the authors performed Raman measurements using ultra-low frequency (ULF) filters on lysozyme and Na^+/K^+-ATPase powder samples and found some strong peaks around 0.8 THz [3, 7]. Besides the detection of terahertz vibrations, spectroscopy techniques were also applied to investigate protein conformational changes [8]. For this purpose, THz-TDS (terahertz time-domain spectroscopy) turned out to be a promising tool in monitoring protein shape changes [9, 10].

From a numerical perspective, several researches correlated the slowest motions to protein conformational changes by means of normal mode calculations [11–13]. In this contribution, we present a similarity analysis between the displacement field of lysine-arginine-ornithine (LAO) binding protein conformational change and its terahertz (expansion-contraction) vibrational modes, evaluated by means of the coarse-grained mechanical model developed in [5]. The aim was to confirm that also expansion-contraction low-frequency vibrations strongly contribute to protein conformational change, as well as to investigate the involved frequency range.

D. Scaramozzino (✉) · G. Lacidogna · A. Carpinteri
Politecnico di Torino, Department of Structural, Geotechnical and Building Engineering, Torino, Italy
e-mail: domenico.scaramozzino@polito.it

M. E. Grady (ed.), *Mechanics of Biological Systems and Materials & Micro-and Nanomechanics, Volume 4*,
Conference Proceedings of the Society for Experimental Mechanics Series, https://doi.org/10.1007/978-3-030-30013-5_2

Methodology

The analysis was conducted by means of three similarity indexes. The first one (*SI*) is defined by the following equation:

$$SI_i = \frac{| \boldsymbol{\delta}_i{}^T \boldsymbol{CC} |}{\sqrt{\boldsymbol{\delta}_i{}^T \boldsymbol{\delta}_i} \sqrt{\boldsymbol{CC}^T \boldsymbol{CC}}}, \quad (1)$$

where $\boldsymbol{\delta}_i$ and $\boldsymbol{CC}$ represent the ith vibrational mode and the conformational change displacement field, respectively. Both of them are *3N*-vectors, being *N* the number of protein residues, i.e., C_α atoms. The former was evaluated by means of modal analysis on protein coarse-grained model, as described in [5]; the latter was computed by vector difference of the two protein reference states, after they have been superimposed. In the remainder of the text, the two protein conformations will be referred to as "open state" and "closed state". By applying Eq. (1), one can evaluate a numerical estimate of the contribution of the ith mode to the conformational change. In particular, *SI* is equal to 1 if $\boldsymbol{\delta}_i$ is equal $\boldsymbol{CC}$, whereas it is equal to 0 if the vectors are orthogonal. Another coefficient was then introduced, i.e. the cumulative-squared similarity index (*CSSI*), defined by the following formula:

$$CSSI_i = \sum_{k=1}^{i} SI_k{}^2, \quad (2)$$

which aims at evaluating the contribution of the first *i* vibrational modes to the conformational change. Considering that the eigenmodes constitute a normal vector basis, by simple calculations, one obtains $CSSI_{3N} = 1$. Note that, if all the vibrational modes had the same contribution to the conformational change displacement field, i.e. $SI_i = SI_{const}$, one would obtain:

$$CSSI_{3N} = 1 = \sum_{k=1}^{3N} SI_k{}^2 = \sum_{k=1}^{3N} SI_{const}{}^2 \Rightarrow SI_{const}{}^2 = \frac{1}{3N}. \quad (3)$$

Therefore, a third similarity index can be defined, namely the normalized-squared similarity index (*NSSI*), which is defined as follows:

$$NSSI_i = \frac{SI_i{}^2}{SI_{const}{}^2}, \quad (4)$$

and provides a numerical estimate of the relative contribution of the ith vibrational mode to the conformational change, with respect to the case in which all the modes had the same involvement.

Results and Discussion

As a case study, lysine-arginine-ornithine (LAO) binding protein was selected, a 238-residues molecule. The reference open and closed states (Fig. 1) were taken from Protein Data Bank [14] (pdb codes: 2lao and 1lst, respectively).

After performing modal analysis on both coarse-grained structures, the similarity analysis was conducted for both the open-closed and closed-open transition (Figs. 2 and 3), and approximately the same results were obtained for both conformational changes. Coherently, it was found that the first six vibrational modes, which refer to rigid motions at zero-frequency, exhibit no contribution to the conformational change. The most involved expansion-contraction vibrational mode is the seventh one for both transitions, leading to $SI_7 = 0.41$ and $SI_7 = 0.46$, for the open-closed and closed-open change, respectively. The similarity index then decreases rapidly for higher mode numbers (Figs. 2a and 3a). As far as the cumulative index distribution is concerned, it can be noted that *CSSI* increases sharply for low mode numbers and then it advances more slightly (Figs. 2b and 3b). Finally, as can be seen from the normalized-squared similarity index distribution, very high values (up to 120–150) are found for the lowest modes, whereas it approaches to zero for higher mode numbers (Fig. 2c and 3c).

Although from a theoretical point of view, all vibrational modes are needed to define any displacement field, in this case, the first 100 ones (on a total of more than 700) are able to describe almost 90% of protein conformational change (Figs. 2b and 3b), thus confirming that also expansion-contraction low-frequency motions contribute more than high-frequency ones.

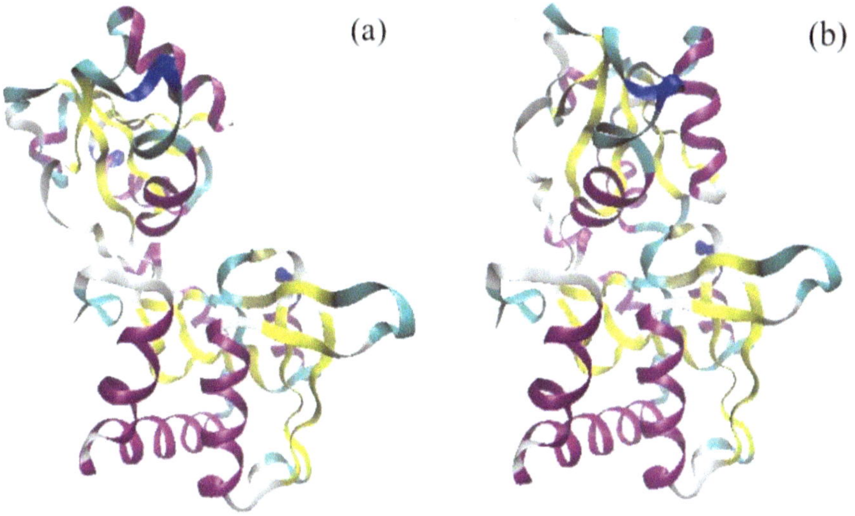

Fig. 1 LAO binding protein: (**a**) open state, (**b**) closed state. Figures obtained by VMD [15]

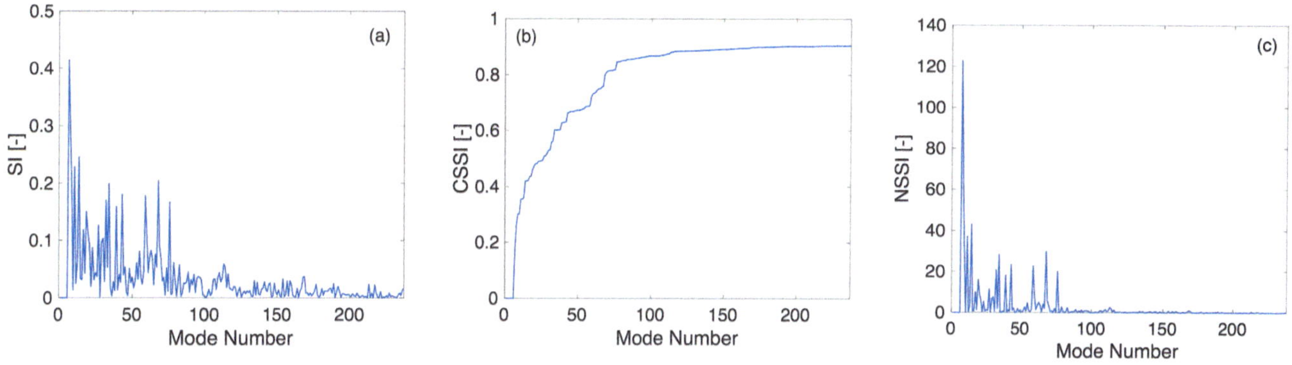

Fig. 2 Similarity indexes (open-closed conformational change): (**a**) *SI*, (**b**) *CSSI*, (**c**) *NSSI*

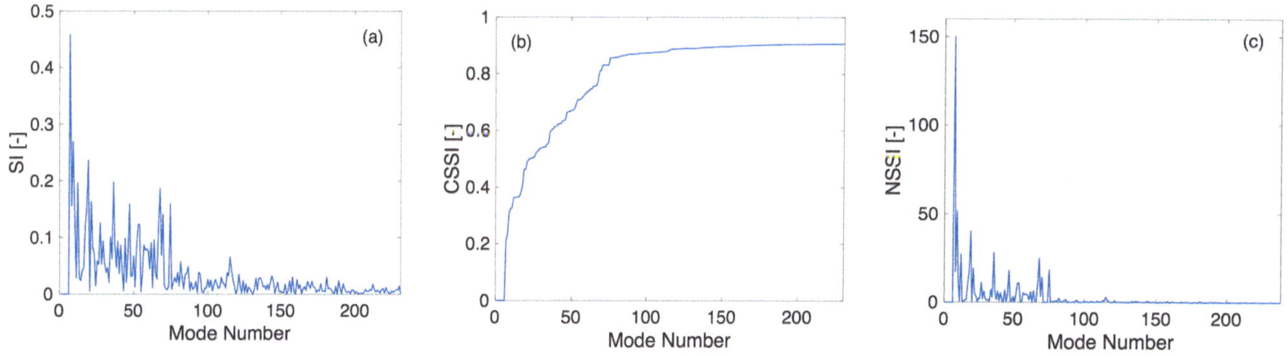

Fig. 3 Similarity indexes (closed-open conformational change): (**a**) *SI*, (**b**) *CSSI*, (**c**) *NSSI*

Finally, by associating each mode number to the corresponding frequency of vibration, one can observe that most of the conformational change occurs within the frequency range around and below 1 THz (~0.1–2 THz), which is the range in which *NSSI* shows very high values (Fig. 4).

It must be noted that the obtained similarity indexes are not so high, the maximum values being lower than 0.5. Besides the fact that only the expansion-contraction modes of the protein backbone are described here, other explanations could be provided for such results. First, modal analysis calculations deal with the evaluation of vibrations around the equilibrium position of the protein structure, i.e., under the assumption of small deformations and linear elasticity; however, the conformational transition may also imply large displacements which, in turn, can be associated with some (geometrical)

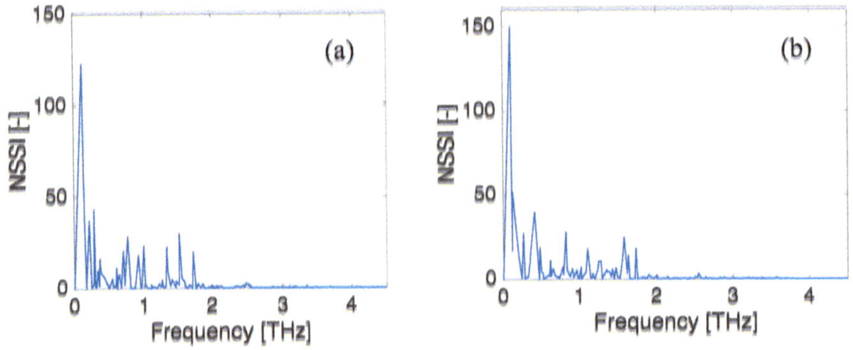

Fig. 4 *NSSI* vs vibrational frequencies: (**a**) open-closed, (**b**) closed-open conformational change

nonlinearities. Secondly, when evaluating the displacement field of the shape change as the vector difference between the two superimposed protein structures, one is implicitly assuming that it happens with a linear transition from the initial state to the final one; however, it is likely that the actual conformational change implies curve pathways as well.

Conclusions

In this contribution, we presented a similarity analysis between the displacement field of LAO binding protein conformational change and the expansion-contraction low-frequency (THz) vibrational modes, obtained via modal analysis by means of a simplified coarse-grained model. Numerical results confirmed that low-frequency motions are the most involved within the protein conformational change and, according to the developed mechanical model, it was found that the protein transition exhibits strong fingerprints within the frequency range between 0.1 and 2 THz. It is strongly believed by the authors that further experimental researches focusing on THz frequencies, for example by means of THz-TDS and low-frequency Raman spectroscopy, could provide interesting insights on the mechanisms underlying protein biological functionality.

References

1. S.E. Lee, R.D. Kamm, M.R.K. Mofrad, A molecular perspective on mechanotransduction in focal adhesion, in *Cellular mechanotransduction: diverse perspectives from molecules to tissues*, ed. by M. R. K. Mofrad, R. D. Kamm, (Cambridge University Press, Cambridge, 2009)
2. S. Mahajan, Y.H. Sanejouand, On the relationship between low-frequency normal modes and large-scale conformational changes of proteins. Arch. Biochem. Biophys. **567**, 59–65 (2015)
3. A. Carpinteri, G. Lacidogna, G. Piana, A. Bassani, Terahertz mechanical vibrations in lysozyme: Raman spectroscopy vs modal analysis. J. Mol. Struct. **1139**, 222–230 (2017)
4. A. Carpinteri, G. Piana, A. Bassani, G. Lacidogna, Terahertz vibrational modes in Na/K-ATPase. J. Biomol. Struct. Dyn. **37**, 256–264 (2019)
5. G. Lacidogna, D. Scaramozzino, G. Piana, A. Carpinteri, Terahertz protein vibrations: The usefulness of coarse-grained numerical models. in: M.E. Grady (ed.), Mechanics of Biological Systems and Materials & Micro- and Nanomechanics, Volume 4, Conference Proceedings of the Society for Experimental Mechanics Series. https://doi.org/10.1007/978-3-030-30013-5_1
6. K.G. Brown, S.C. Erfurth, E.W. Small, W.L. Peticolas, Conformationally dependent low-frequency motions of proteins by laser Raman spectroscopy. Proc. Natl. Acad. Sci. U. S. A. **69**, 1467–1469 (1972)
7. G. Lacidogna, G. Piana, A. Bassani, A. Carpinteri, Raman spectroscopy of Na/K-ATPase with special focus on low-frequency vibrations. Vib. Spectrosc. **92**, 298–301 (2017)
8. D.A. Turton, H.M. Senn, T. Harwood, A.J. Lapthorn, E.M. Ellis, K. Wynne, Terahertz underdamped vibrational motion governs protein-ligand binding in solution. Nat. Commun. (2014). https://doi.org/10.1038/ncomms4999
9. E. Castro-Camus, M.B. Johnston, Conformational changes of photoactive yellow protein monitored by terahertz spectroscopy. Chem. Phys. Lett. **455**, 289–292 (2008)
10. H. Chen, G.Y. Chen, S.Q. Li, L. Wang, Reversible conformational changes of PsbO protein detected by terahertz time-domain spectroscopy. Chin. Phys. Lett. **26**, 084204 (2009)
11. F. Tama, Y.H. Sanejouand, Conformational change of proteins arising from normal mode calculations. Protein Eng. **14**, 1–6 (2001)
12. W. Zheng, B.R. Brooks, Normal-mode-based prediction of protein conformational changes guided by distance constraints. Biophys. J. **88**, 3109–3117 (2005)
13. P. Petrone, V.S. Pande, Can conformational change be described by only a few normal modes? Biophys. J. **90**, 1583–1593 (2006)
14. Protein Data Bank. https://www.rcsb.org
15. W. Humphrey, A. Dalke, K. Schulten, VMD–visual molecular dynamics. J. Mol. Graph. **14**, 33–38 (1996)

Residual Stresses in Biological Materials

Herbert Silva and Drew Nelson

Abstract Examples of methods for determining residual stresses and strains (RSS) in biological materials are reviewed and postulated roles of RSS in biomechanical behavior described. Residual strains are thought to exert a particularly important influence on the behavior of arteries. For several decades, determination of those strains has relied on the opening angle method, in which a ring removed from an artery is slit radially, causing the ring to spring open. The change in geometry of the ring provides input to analytical relations for estimating circumferential residual strains. The wall of an artery, which has three layers, contains a mixture of elastin, collagen fibrils and muscle cells. An attempt to use small angle X-ray scattering (SAXS) to characterize residual strains using collagen fibrils as internal "micro strain sensors" is presented. First, the results of SAXS experiments to investigate the response of collagen fibrils to strains applied to arterial tissue are presented. Strains as measured in fibrils are compared to those applied to the tissue. Then, SAXS experiments to explore residual strains in collagen fibrils within rings of arterial tissue are described. Results are compared to tissue-level residual strains estimated from the opening angle method.

Keywords Artery · Collagen · Diffraction · X-ray · Residual stress

Introduction

Methods for determining residual stresses in engineering materials have been the subject of numerous publications over the years. Similar methods have been used to study residual stresses and strains (RSS) in biological materials. Two examples follow.

Opening Angle Method

An approach for estimating circumferential residual stresses in thin-walled tubing involves slitting a length of tube longitudinally, causing it to spring open upon release of residual stresses, as in Fig. 1a. Circumferential residual stresses assumed to vary linearly through the wall can then be computed [1].

A similar experimental approach has been applied to arteries, using constitutive relations appropriate for biological tissue. Fung [2] and Vaishnav and Vossoughi [3] found that when a short length of an artery was sliced to produce a ring and then slit radially through the wall, it became a sector like that depicted in idealized form in Fig. 1b. Chuong & Fung [4] obtained an analytical solution for circumferential residual strains based on measurement of the opening angle and associated radii.

Strains were found to be compressive at inner diameters of arteries. Incorporating circumferential residual strains in the computation of circumferential stresses fromblood pressure is predicted to reduce stresses at the inner diameter significantly, as seen in Fig. 2.

By counteracting stresses from blood pressure in arteries [6–9], compressive circumferential RSS may increase resistance to failure [10] (e.g., aortic tearing). If high blood pressure develops, RSS have been found to increase and help reduce stresses at the inner diameter of arteries [11, 12]. In addition to such physiological roles, RSS may also be of clinical significance. For instance, a longitudinal incision in an artery will cause newly created surfaces to spring open. RSS governs the size and shape of openings and are important in computational simulations of coronary bypass surgery [13] and angioplasty [14].

H. Silva · D. Nelson (✉)
Mechanical Engineering Department, Stanford University, Stanford, CA, USA
e-mail: herbertp@stanford.edu; dnelson@stanford.edu

M. E. Grady (ed.), *Mechanics of Biological Systems and Materials & Micro-and Nanomechanics, Volume 4*,
Conference Proceedings of the Society for Experimental Mechanics Series, https://doi.org/10.1007/978-3-030-30013-5_3

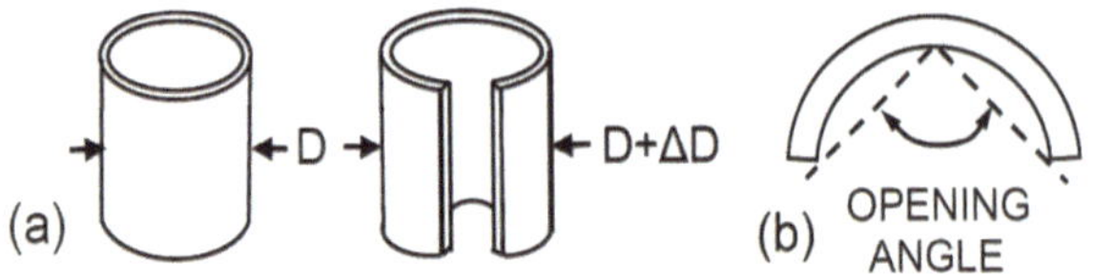

Fig. 1 (**a**) Change in diameter after slitting to release circumferential residual stress and (**b**) deflection of an arterial ring after slitting

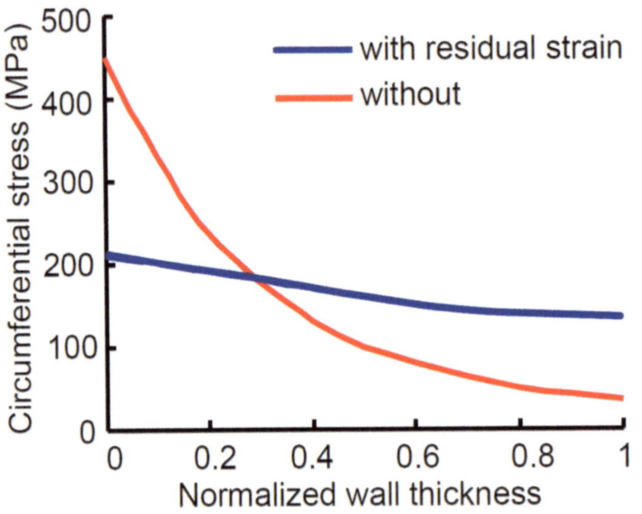

Fig. 2 Effect of residual strains on stresses computed from blood pressure, based on data from [5]

The opening angle method has also been applied to study RSS in veins [15], the esophagus [16–19], intestine [20, 21], ventricle of the heart [22–25], tissue-engineered heart chambers [26], trachea [27], airways in lungs [28], brain [29], and an intervertebral (spinal) disc [30]. Shortcomings of the method are discussed, for example, in Refs. [7, 31].

X-Ray Diffraction

The measurement of residual stresses by X-ray diffraction (XRD) is well-established for engineering materials with crystalline structures [32]. Bone contains crystals of the mineral bioapatite, to which XRD can be applied. For instance, Tung et al. [33] performed experiments using small samples extracted from a bovine femur. The specimen geometry is shown schematically in Fig. 3a. Initially, residual stresses in four hydrated specimens were compressive, as shown in Fig. 3b. When specimens were allowed to dehydrate, the compressive residual stresses diminished and became tensile. Residual stresses (tensile) were also observed in collagen fibrils present in bone using small angle X-ray scattering, a technique summarized in the next section. A review of other experiments using XRD to study residual stresses in bone is provided by Tadano & Yamaha [34].

RSS can play an important role in simulations of bone remodeling [35]. (Remodeling refers to a reorganization or renovation of tissue.) RSS may influence bone healing [36] and the morphogenesis of limbs [37]. The change in RSS with dehydration in Fig. 3b provides an example of how sensitive RSS measured in biological tissues can be to experimental conditions.

Small Angle X-Ray Scattering Overview

In crystalline materials, constructive interference of diffracted X-rays will occur at certain angles θ_{hkl} that depend on the spacing d_{hkl} of a set of diffracting plane designated by indices (hkl). A relation between θ_{hkl} and d_{hkl} is given by Bragg's law [38]:

$$n\lambda = 2d_{hkl} \sin \Theta_{hkl} \tag{1}$$

where λ = X-ray wavelength and n = integral order.

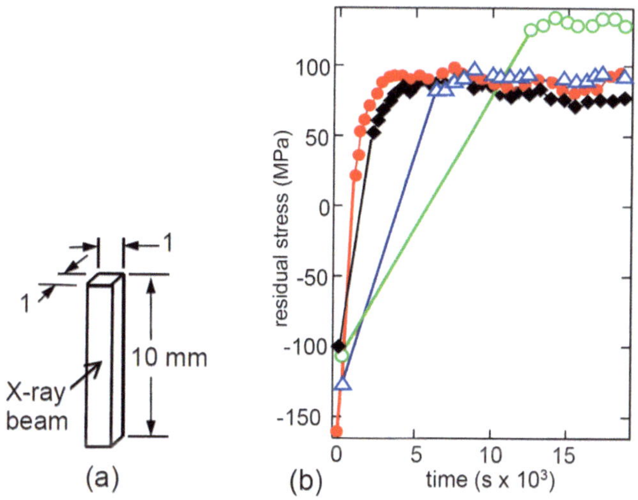

Fig. 3 (**a**) Geometry of bone specimens and (**b**) residual stress in bioapatite crystals in four specimens dehydrating in air, re-plotted from data in [33]

Fig. 4 Schematic of a transmission XRD setup and an image of diffraction rings observed using a 2D detector

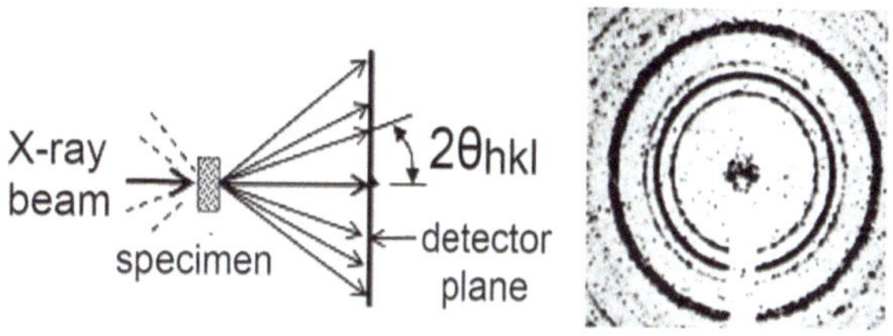

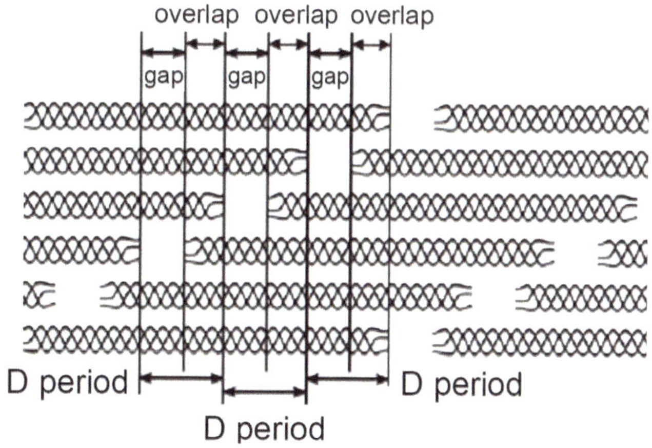

Fig. 5 Schematic of a collagen fibril showing D-period

If enough crystals are favorably oriented to provide constructive interference for a given θ_{hkl}, diffracted X-rays will form a cone [39]. Diffraction cones will produce rings on a detector plane as illustrated in Fig. 4. A 2D detector can monitor the number of diffracted X-ray photons per second reaching a given location on the detector.

Small angle X-ray scattering (SAXS) can obtain structural information about materials with a periodic feature D that is much larger than d_{hkl}, requiring measurements at smaller angles θ than in XRD. In the experiments to be described in the next section, collagen fibrils were the material of interest for SAXS. Three strands of polypeptide twisted into a helix about 3000 Å long comprise the basic structural unit of a collagen molecule [40]. The molecules self-assemble into collagen fibrils with a repeating longitudinal feature D caused by staggering of collagen molecules within a fibril, as seen in Fig. 5. The length of D is known as the "D period" and is typically on the order of 680 Å.

It is common in SAXS to plot data in terms of the scattering parameter $Q = (4\pi \sin \theta)/\lambda$. From Eq. 1, the parameter Q can also be expressed as $Q = 2\pi n/D$. Scattering data from collagen fibrils can be processed to find the variation of intensity with Q, producing results such as shown in Fig. 6a. The periodic pattern of peaks corresponds to different n values in Eq. 1, with d_{hkl} replaced by the D-period.

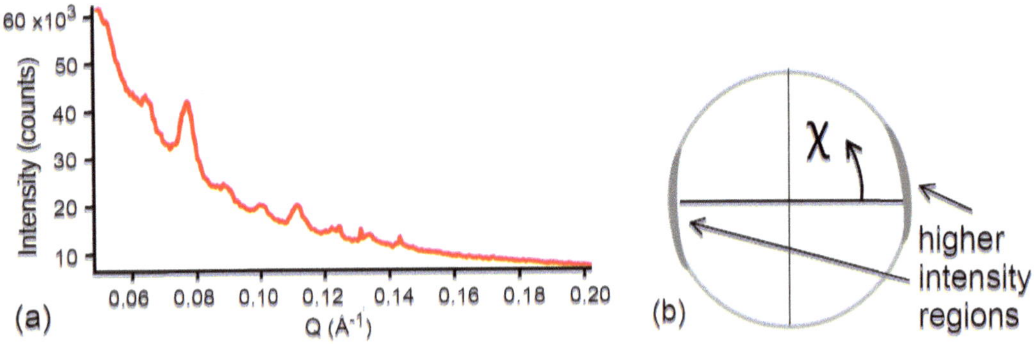

Fig. 6 (**a**) Example of scattering peaks for collagen fibrils and (**b**) schematic of intensity variation around a ring from oriented collagen fibril scattering obtained from a 2D detector

Fig. 7 Layers of an artery. The intima is a thin cellular layer. The media contains elastin and collagen as well as muscle cells. The adventitia primarily contains connective tissue

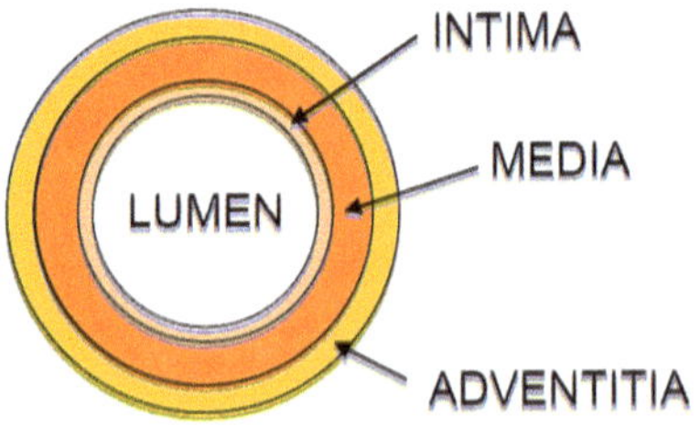

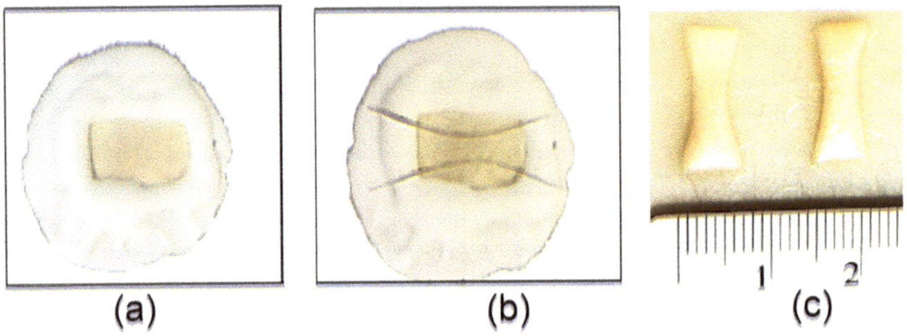

Fig. 8 (**a**) Tissue specimen surrounded by frozen OCT, (**b**) specimen shape after use of a cookie cutter device and (**c**) specimens after thawing of OCT, with numbers representing cm

When X-rays are scattered from collagen fibrils with a preferred orientation, arcs will appear at a 2D detector as in Fig. 6b. By determining the variation of intensity around a ring as a function of angle χ, it is possible to obtain information about fiber orientation [41].

Experiments

The goals of the experiments were to: (a) perform stretching tests with arterial tissue specimens to explore collagen fibril strains found by SAXS and (b) observe how fibril strains might change when tissue-level (macroscopic) residual strains were released from ring specimens of an artery by slitting, as in Fig. 1b.

A fresh porcine aorta was obtained from a medical research supplier. A specimen was cut from the arterial wall and frozen flat using an optimal cutting temperature (OCT) compound. The intimal layer (Fig. 7) of the specimen contacted a metal plate. A microtome was used to remove the adventitial layer (Fig. 7) and produce a uniform specimen thickness of 1 mm. Next, the specimen was cut into rectangular pieces (Fig. 8a) with longitudinal axes aligned with the circumferential direction of the artery. Specimens with dog bone shapes (Fig. 8b) were formed using a custom made device similar to a cookie cutter.

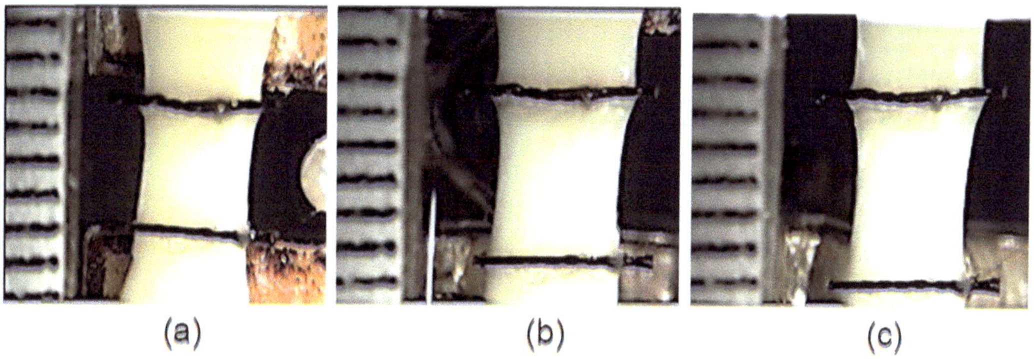

Fig. 9 Images of tissue specimen with applied displacements of (a) zero, (b) 1 and (c) 2 mm

Fig. 10 (a) Removal of a ring specimen from aorta and (b) regions scanned to obtain X-ray scattering data from collagen fibrils

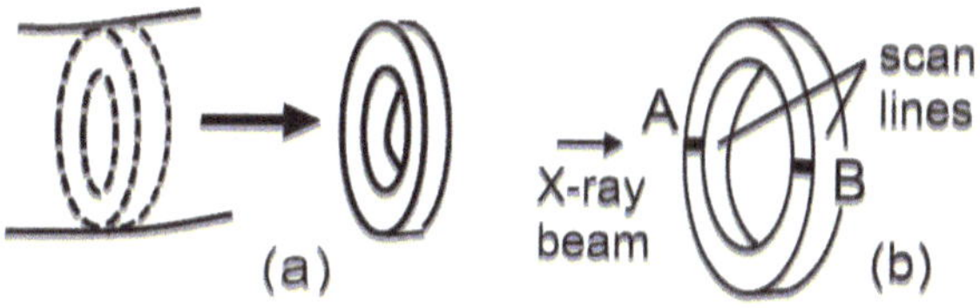

Fig. 11 (a) Ring just prior to scanning and (b) after a radial cut (with mm scales shown)

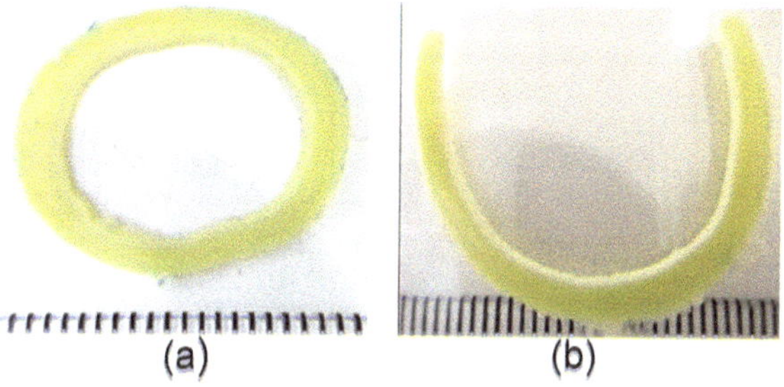

Just prior to testing, two short lengths of fine black sewing thread were coated with cyanoacrylate adhesive and attached to tissue specimens to serve as markers, as seen in Fig. 9, for monitoring changes in displacement of "gage sections" during stretching. Then a specimen was attached to the grips of a stretching device, also using cyanoacrylate. The grips had fine sand paper to help keep a specimen attached.

Experiments were performed using synchrotron X-rays with a beam normal to the plane of a specimen shown in Fig. 9. The beam size was 300×400 μm. At each level of applied displacement, SAXS data were obtained along the centerline a specimen in the middle of the gage section. Specimens were kept moist during testing by a setup that allowed water to drip onto a specimen approximately every 20 s. A similar type of approach was used by Gupta et al. [42]. The SAXS data from each specimen were processed to produce plots of intensity vs. Q and angle χ (Fig. 6).

Residual strain experiments used ring specimens. As depicted in Fig. 10a, short lengths of a porcine aorta were removed by slicing, resulting in rings that had a nearly circular cross-section and a thickness of approximately 2.5 mm.

For X-ray scattering experiments, the rings were enclosed in thin Kapton film to preserve hydration. Two adjacent rings were taken from near the top of the aorta (closer to the heart). An image of each ring was taken prior to enclosure in Kapton. Rings were then scanned along horizontal lines shown in Fig. 10b. After SAXS data were obtained, a radial cut was made in each of the rings to release tissue-level residual strains, causing the rings to spring open as in Fig. 11. Then an image of each opened ring was taken. Next, the small regions that had been scanned were removed from the rings and re-scanned to gather SAXS data for comparison with data when a ring was intact. Data were collected with a beam size of 250×250 μm at eight locations from the OD to ID along scan line A for ring 1 and twelve along line B from the ID to OD. For ring 2, line A had ten locations and line B had eleven.

To estimate tissue residual strains, the inner and outer perimeters of a ring before and after opening were measured from images such as those in Fig. 11 using Image J software [43]. With those measurements as an input, circumferential residual strains were estimated using the approach given by Fung & Liu [44]. The scattering data were analyzed to produce plots of intensity vs. parameter Q.

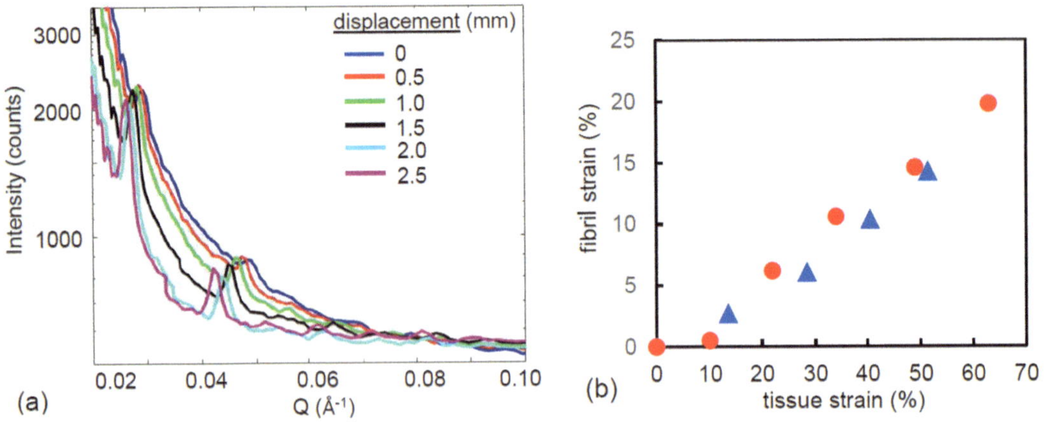

Fig. 12 (**a**) Change in collagen fibril peak location with stretching and (**b**) relation between fibril strain and tissue strain for two specimens

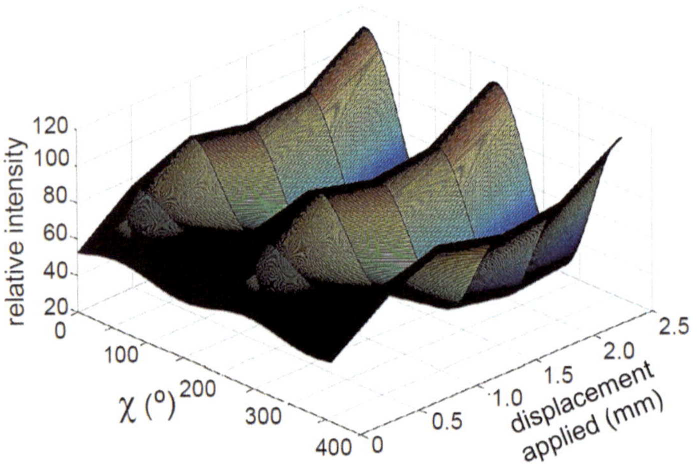

Fig. 13 Intensity vs. angle χ and applied displacement, showing preferred orientation. The stretching direction is at $\chi = 0°$ ($=180° = 360°$)

Results

In the stretching experiments, a shift in locations of collagen fibril peaks with increased stretching occurred, as seen in Fig. 12a for one of the specimens. Tissue-level strains were computed as the change in distance between specimen markers divided by the original distance. The fibril strains at the middle of a specimen were computed from the changes in spacing between the third and fifth order peaks divided by the spacing at no applied displacement. Those peaks were used since the other peaks in Fig. 12 did not have a sufficient intensity. Figure 12b shows a relation between local collagen fibril strain and tissue strain. The fibril strains were a fraction of the tissue level strains, consistent with observations in SAXS experiments with other cardiovascular tissues [45, 46].

The I(χ) profiles from the stretching experiments showed collagen fibrils to be aligned with the stretching direction, which also corresponded to the circumferential direction of the porcine aorta. The degree of orientation increased as stretching increased, as seen in Fig.13. The use of collagen fibrils as tissue "strain sensors" in the stretching experiments offered reasonable hope that the fibrils might provide information about residual strains in tissues.

Slitting experiments revealed circumferential residual strains of approximately -12% (compressive) at the ID of ring specimens tested, decreasing towards the OD. However, corresponding shifts in collagen fibril peaks on I vs. Q plots were minimal, as shown for example in Fig. 14. Shifts of peaks from the numerous locations scanned were within a band of $\pm 2\%$ of zero with no trend apparent. Reasons for lack of significant fibril response are currently unknown.

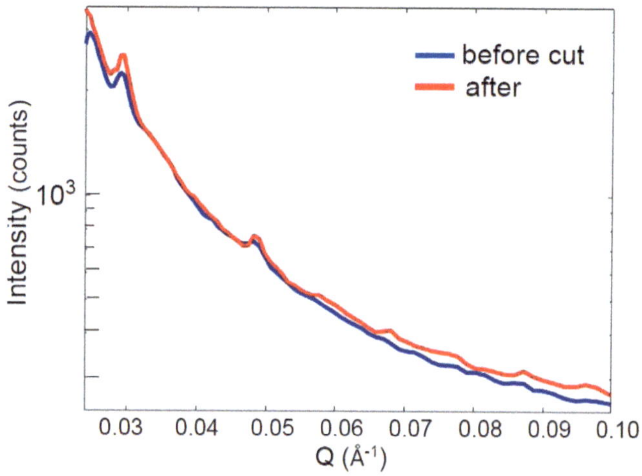

Fig. 14 Representative plot of intensity vs. parameter Q before and after residual stress release, near the inner diameter of a ring

Conclusions

1. Collagen fibril strains in specimens of porcine aorta tissue were measured successfully by small angle X-ray scattering (SAXS) in stretching tests.
2. Collagen fibrils became increasingly oriented with the direction of stretching (also the circumferential direction in aorta) as displacements applied to specimens increased.
3. Significant changes in the D-period of collagen fibrils were not observed upon release of tissue-level residual strains in rings taken from an aorta.

References

1. *Standard practice for estimating the approximate residual circumferential stress in straight thin-walled tubing, E1928–07* (ASTM, West Conshohocken, PA)
2. Y. Fung, *Biodynamics: circulation* (Springer-Verlag, New York, 1984), pp. 54–60
3. R. Vishnav, J. Vossoughi, Estimation of residual strains in aortic segments, in *Biomedical engineering II, recent developments*, ed. by C. Hall, (1983), pp. 330–333
4. C. Choung, Y. Fung, On residual stresses in arteries. J. Biomech. Eng. **108**, 189–192 (1986)
5. W. Zhang et al., The effect of longitudinal pre-stress and radial constraint on the stress distribution in the vessel wall: a new hypothesis. Mol. Cell. Biomech. **2**, 41–52 (2005)
6. Y. Fung, What are the residual stresses doing in our blood vessels? Ann. Biomed. Eng. **19**, 237–249 (1991)
7. A. Rachev, S. Greenwald, Residual strains in conduit arteries. J. Biomech. **36**, 661–670 (2003)
8. M. Destrade et al., Uniform transmural strain in pre-stresses arteries occurs at physiological pressure. J. Theor. Biol. **303**, 93–97 (2012)
9. A. Delfino et al., Residual strain effects on the stress field in a thick wall finite element model of the human carotid bifurcation. J. Biomech. **30**, 777–786 (1997)
10. K. Volokh, Prediction of arterial failure based on microstructural bi-layer fiber-matrix model with softening. J. Biomech. **41**, 447–453 (2008)
11. C. Wang, G. Kassab, Increase in opening angle in hypertension off-loads the intimal stress: a simulation study. J. Biomech. Eng. **131**, 114502 (2009). https://doi.org/10.1115/1.4000085
12. Y. Fung, S. Liu, Change in zero-stress state of rat pulmonary arteries in hypoxic hypertension. J. Appl. Physiol. **70**, 2455–2424 (1991)
13. F. Cacho et al., A procedure to simulate coronary artery bypass graft surgery. Med. Biol. Eng. Comput. **45**, 819–827 (2007)
14. X. Zhao et al., A novel arterial constitutive model in a commercial finite element package: application to balloon angioplasty. J. Theor. Biol. **286**, 92–99 (2011)
15. J. Xie et al., The zero-stress state of rat veins. J. Biomech. Eng. **113**, 36–41 (1991)
16. H. Gregersen et al., Strain distribution in the layered wall of the esophagus. J. Biomech. Eng. **121**, 442–448 (1999)
17. D. Laio et al., Stress distribution in the layered wall of the rat oesophagus. Med. Eng. Phys. **25**, 731–738 (2003)
18. J. Zhao et al., Opening angle and residual strain in a three-layered model of pig oesophagus. J. Biomech. **40**, 3187–3192 (2007)
19. D. Sokolis, Strain-energy function and three-dimensional stress distribution in esophageal biomechanics. J. Biomech. **43**, 2753–2764 (2010)
20. C. Gao, H. Gregersen, Biomechanical and morphological properties in rat large intestine. J. Biomech. **33**, 1089–1097 (2000)
21. Y. Dou et al., Longitudinal residual strain and stress-strain relationship in rat small intestine. Biomed. Eng. Online **5**, 37 (2006). https://doi.org/10.1186/1475-925-5-37

22. L. Taber et al., Residual strain in the ventricle of the stage 16-24 chick embryo. Circ. Res. **72**, 455–462 (1993)
23. J. Omens, Y. Fung, Residual strain in rat left ventricle. Circ. Res. **66**, 37–45 (1990)
24. S. Summerour et al., Residual strain in ischemic ventricular myocardium. J. Biomech. Eng. **120**, 710–714 (1998)
25. J. Omens et al., Complex distribution of residual stress and strain in the mouse left ventricle: experimental and theoretical models. Biomech. Model. Mechanobiol. **1**, 267–277 (2003)
26. E. Lee et al., Engineered cardiac organoid chambers: toward functional biological model ventricle. Tissue Eng. Part A **14**, 215–225 (2007)
27. H. Han, Y. Fung, Residual strains in porcine and canine trachea. J. Biomech. **24**, 307–315 (1991)
28. K. McKay et al., Zero-stress state of intra- and extraparenchymal airways from human, pig, rabbit and sheep lungs. J. Appl. Physiol. **92**, 1261–1266 (2002)
29. G. Xu et al., Residual stress in the adult mouse brain. Biomech. Model. Mechanobiol. **8**, 253–262 (2009)
30. A. Michalek et al., Large residual strains are present in the intervertebral disc annulus fibrosus in the unloaded state. J. Biomech. **45**, 1227–1231 (2012)
31. L. Taber, J. Humphrey, Stress-modulated growth, residual stress and heterogeneity. J. Biomech. Eng. **123**, 528–535 (2001)
32. I. Noyan, J. Cohen, *Residual stress: measurement by diffraction and interpretation* (Springer, New York, 1987)
33. P. Tung et al., Hydration and radiation effects on the residual stress state of cortical bone. Acta Biomater. **9**, 9503–9507 (2013)
34. S. Tadano, S. Yamada, How is residual stress/strain detected in bone tissue? Bull. JSME **3**, 15–00291 (2016)
35. M. Tanaka et al., Mechanical remodeling of bone structure considering residual stress. JSME Int. J. A **39**, 297–305 (1996)
36. L. Gonzalez-Torres et al., Evaluation of residual stresses due to bone callus growth: a computational study. J. Biomech. **44**, 1782–1787 (2011)
37. X. Henderson, D. Carter, Mechanical induction in limb morphogenesis: the role of growth-generated strains and pressures. Bone **11**, 645–653 (2002)
38. R. Dinnebier, S. Billinge, *Powder diffraction: theory and practice* (Royal Society of Chemistry Publishing, Cambridge, UK, 2008)
39. B. He, *Two-dimensional X-Ray diffraction* (John Wiley & Sons, Inc., Hoboken, NJ, 2009)
40. I. Streeter, N. de Leeuw, A molecular dynamics study of the interprotein interactions in collagen fibrils. Soft Matter **7**, 3373–2282 (2011)
41. S. Pabisch et al., Imaging the nanostructure of bone and dentin through small- and wide-angle X-ray scattering. Methods Enzymol. **532**, 391–413 (2013)
42. H. Gupta et al., Cooperative deformation of mineral and collagen in bone at the nanoscale. Proc. Natl. Acad. Sci. **103**, 17741–17746 (2006)
43. https://imagej.nih.gov/ij/
44. Y. Fung, S. Liu, Strain distribution in small blood vessels with zero-stress state taken into consideration. Am. J. Physiol. **262**, H544–H552 (1992)
45. J. Liao et al., Molecular orientation of collagen in intact planar connective tissues under biaxial stretch. Acta Biomater. **1**, 45–54 (2005)
46. F. Schmid et al., In situ tensile testing of human aortas by time resolved small-angle X-ray scattering. J. Synchrotron Radiat. **12**, 727–733 (2005)

Quantification of Papillary Muscle Motion and Mitral Regurgitation After Myocardial Infarction

Connor R. Ferguson, Robert C. Gorman, and Jonathan F. Wenk

Abstract Change in papillary muscle motion as a result of left ventricular (LV) remodeling after posterolateral myocardial infarction is thought to contribute to ischemic mitral regurgitation. A finite element (FE) model of the LV was created from magnetic resonance images acquired immediately before myocardial infarction and 8 weeks later in a cohort of 12 sheep. Severity of mitral regurgitation was rated by two-dimensional echocardiography and regurgitant volume was estimated using MRI. Of the cohort, six animals (DC) received hydrogel injection therapy shown to limit ventricular remodeling after myocardial infarction (Rodell, Christopher B., Circ. Cardiovasc. Interv. 9:e004058 2016) while the control group (MI) received a similar pattern of saline injections. LV pressure was determined by direct invasive measurement and volume was estimated from MRI. FE models of the LV for each animal included both healthy and infarct tissue regions as well as a simulated hydrogel injection pattern for the DC group. Constitutive model material parameters for each region in the FE model were assigned based on results from previous research. Invasive LV pressure measurements at end diastole and end systole were used as boundary conditions to drive model simulations for each animal. Passive stiffness (C) and active material parameter (T_{max}) were adjusted to match MRI estimations of LV volume at end systole and end diastole. Nodal positions of the chordae tendineae (CT) were determined by measurements obtained from the excised heart of each animal at the terminal time point. Changes in CT nodal displacements between end systole and end diastole at 0- and 8-week time points were used to investigate the potential contribution of changes in papillary muscle motion to the progression of ischemic mitral regurgitation after myocardial infarction. Nodal displacements were broken down into radial, circumferential, and longitudinal components relative to the anatomy of the individual animal model. Model results highlighted an outward radial movement in the infarct region after 8 weeks in untreated animals, while radial direction of motion observed in the treated animal group was preserved relative to baseline. Circumferential displacement decreased in the remote region in the untreated animal group after 8 weeks but was preserved relative to baseline in the treated animal group. MRI estimates of regurgitant volume increased significantly in the untreated animal group after 8 weeks but did not increase in the treated group. The results of this analysis suggest that hydrogel injection treatment may serve to limit changes in papillary muscle motion and severity of mitral regurgitation after posterolateral myocardial infarction.

Keywords Magnetic resonance imaging · Finite element modeling · Displacement · Volume analysis · Mitral regurgitation

Introduction and Background

Myocardial infarction is known to cause left ventricular (LV) remodeling associated with dilatation and distortion of the ventricular shape leading to deterioration in contractile function [1]. A previous population-based study by Bursi et al. identified mitral regurgitation in as many as 50% of patients after the first myocardial infarction [2]. The presence of mitral

C. R. Ferguson (✉)
Department of Mechanical Engineering, University of Kentucky, Lexington, KY, USA
e-mail: crfe222@g.uky.edu

R. C. Gorman
Gorman Cardiovascular Group, Department of Surgery, University of Pennsylvania, Philadelphia, PA, USA
e-mail: robert.gorman@uphs.upenn.edu

J. F. Wenk
Department of Mechanical Engineering, University of Kentucky, Lexington, KY, USA

Department of Surgery, University of Kentucky, Lexington, KY, USA
e-mail: jonathan.wenk@uky.edu

M. E. Grady (ed.), *Mechanics of Biological Systems and Materials & Micro-and Nanomechanics, Volume 4*,
Conference Proceedings of the Society for Experimental Mechanics Series, https://doi.org/10.1007/978-3-030-30013-5_4

regurgitation was found to be an independent predictor of cardiovascular mortality in post-MI patients [3]. Changes in papillary muscle motion as a result of left ventricular (LV) remodeling after posterolateral myocardial infarction is thought to contribute to ischemic mitral regurgitation [4]. This group previously showed that in-vivo injected shear-thinning biomaterial significantly improved regional infarct contractile function and reduced LV remodeling in a cohort of Dorset sheep when compared to untreated controls [5]. The present study uses a similar animal-specific MRI-based FE model to quantify papillary muscle motion before and 8 weeks after MI in the same ovine cohort. This approach allows us to investigate the potential contribution of changes in papillary muscle motion to the progression of ischemic mitral regurgitation after myocardial infarction.

Analysis

The data presented in this study were previously collected as part of a study by Rodell et al. [6] in compliance with the University of Pennsylvania's Institutional Animal Care and Use Committee and in agreement with the National Institute of Health's guidelines for the care and use of laboratory animals (NIH Publication 85-23, revised 1996). In summary, a cohort of Dorset sheep underwent a left thoracotomy and subsequent posterolateral infarction produced by suture ligation of obtuse marginal branches to result in an infarct area of approximately 20% of the LV and involving the posterior papillary. Infarct area coverage of the papillary muscle was unsatisfactory in one animal from each of the untreated and treated groups, and those data are excluded from this analysis. Thirty minutes post ligation, 16 injections (0.3 mL each via ½ mL syringe, with 27G ½" needle) consisting of saline (control) or hydrogel were administered in the infarct region. MRI acquisition was performed at 3T (Tim Magnetom Trio Scanner; Siemens, Inc.) immediately before infarct as well as 8 weeks post infarct. Cardiac gating was performed by placement of a pressure catheter (Millar Instruments, Inc.) into the LV. LV geometry was obtained from 2D CINE images using ImageJ as shown in Fig. 1 and infarct location and geometry was confirmed using additional late gadolinium enhancement imaging. At 8 weeks post-MI, the animals were euthanized and the hearts were excised.

A finite element model of the LV was created from the MRI acquisition for each animal at both 0-week and 8-week time points as shown in Fig. 2.

LV pressure was determined by catheterization and LV volume was estimated from MRI. Severity of mitral regurgitation was rated (scale 0–3) by two-dimensional echocardiography. FE models for each animal included both healthy and infarcted tissue regions as well as a simulated hydrogel injection pattern for the treatment group based on previous MRI data [6]. The passive myocardium response was represented using a nearly incompressible, transversely isotropic, hyperelastic material defined by the following strain energy function [7]:

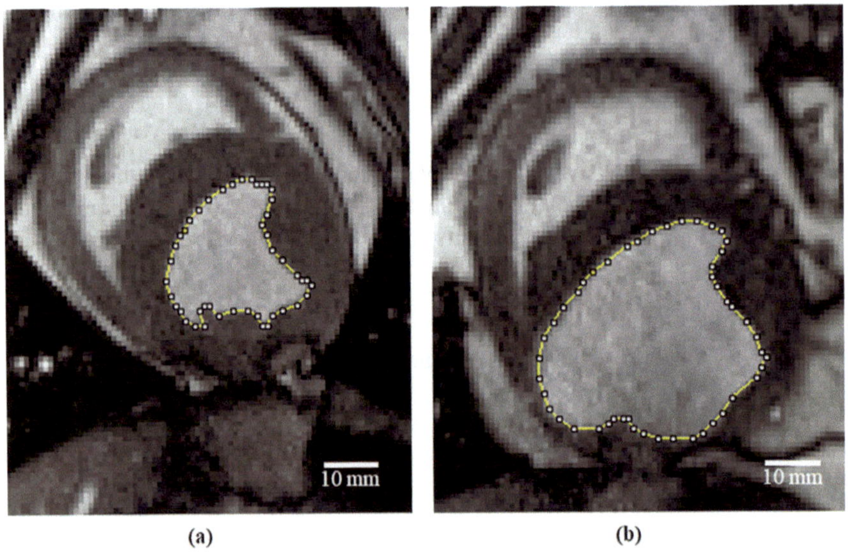

Fig. 1 LV endocardial contours on one short axis slice (**a**) immediately before infarct and (**b**) 8 weeks later in the same untreated animal. Surfaces generated from endocardial and epicardial contours were used to create individual finite element models for each animal

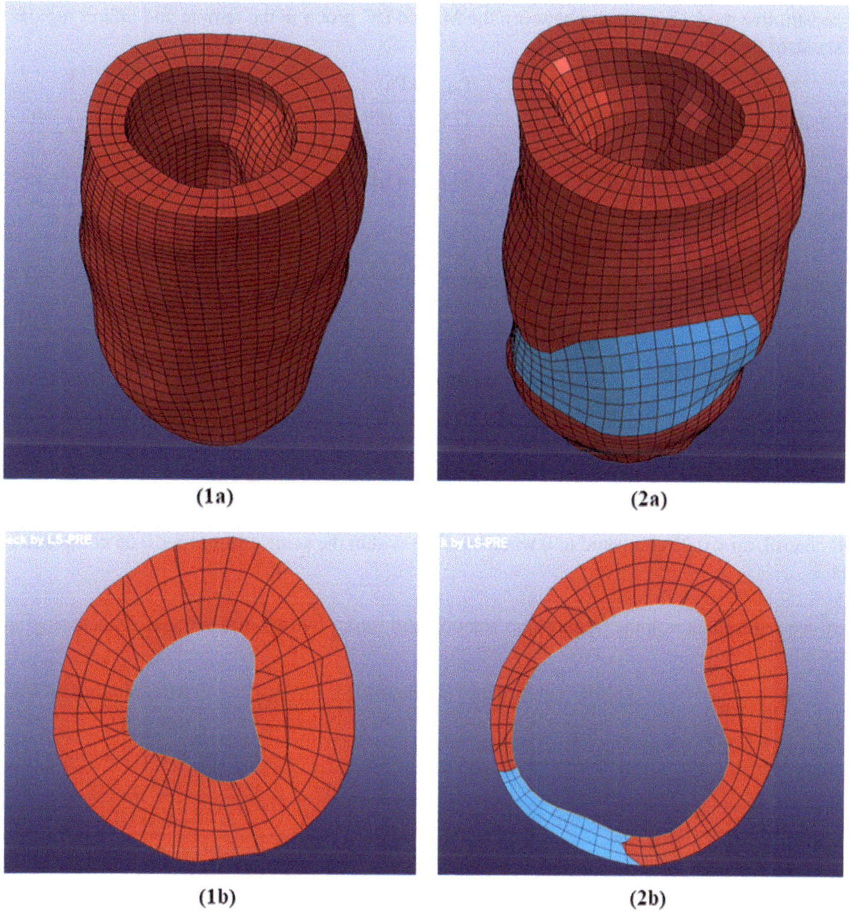

Fig. 2 Full LV FE model (**a**) and FE model short axis slice (**b**) immediately before infarct (1) and 8 weeks later (2) in the same untreated animal

$$W_{myocardium} = \frac{C}{2} \times \left(e^{b_f E_{ff}^2 + b_t\left(E_{ss}^2 + E_{nn}^2 + E_{ns}^2 + E_{sn}^2\right) + b_{fs}\left(E_{fs}^2 + E_{sf}^2 + E_{fn}^2 + E_{nf}^2\right)} - 1 \right) + \frac{k}{2}(J-1)^2 \quad (1)$$

where E_{ij} are the deviatoric components of the Green-Lagrange strain tensor relative to the myofiber coordinate system (f = fiber direction, s = cross-fiber in-plane direction, n = transverse-fiber direction) and J is the determinant of the deformation gradient. The response of the hydrogel injections in the DC group was represented using a nearly incompressible, isotropic, hyperelastic material law defined by the following strain energy function [5]:

$$W_{injection} = \frac{E}{2(1+v)} \mathrm{tr}\left(\mathbf{E}^2\right) + \frac{E}{6(1-2v)} \ln(J)^2 \quad (2)$$

where **E** is the deviatoric Green-Lagrange strain tensor and the material parameters for Young's modulus (E) of the hydrogel were assigned based on experimental results [6]. Constitutive model material parameters for each region in the FE model were assigned based on results from previous studies [5–9]. Invasive LV pressure measurements at end diastole and end systole were used as boundary conditions to drive model simulations for each animal. Passive stiffness (C) and active material parameter (T_{max}) were adjusted to match calculated LV volume from the model to MRI estimations of LV volume at end diastole and end systole. The constitutive model parameters for each group are given in Table 1. Passive material parameters b_f, b_t, and b_{fs} were assigned differently in the remote and infarct region based on results from previous studies [5, 6].

Nodal positions of the chordae tendineae (CT) were determined by measurements obtained from the excised heart of each animal at the terminal time point. A local coordinate system was created for each CT attachment node and nodal displacement from end diastole to end systole was quantified in terms of the local coordinate system for that node. The positive longitudinal direction was defined as the positive vertical axis relative to the FE model and the positive radial direction was defined as the outward vector pointing away from the long axis of the ventricle toward the corresponding CT

Table 1 Comparison of constitutive model parameters between the MI and DC group in the remote and infarct regions at 8-weeks post-MI. Data are presented as mean ± standard deviation where applicable

		C (kPa)	T_{max} (kPa)	b_f	b_t	b_{fs}
MI	Remote	0.75 ± 0.66	278.7 ± 186.4	22.84	3.46	12.00
	Infarct	7.41 ± 6.92	–	15.28	8.32	25.27
DC	Remote	0.09 ± 0.06	93.6 ± 43.1	22.84	3.46	12.00
	Infarct	3.26 ± 3.32	–	139.26	18.55	25.74

Table 2 Left ventricle papillary muscle chordae tendineae attachment point nodal displacements. Data are presented in mm as mean ± standard deviation

		Remote			Infarct		
		Radial	Circumferential	Longitudinal	Radial	Circumferential	Longitudinal
MI	0 week	−5.2 ± 1.9	−3.2 ± 0.5	1.8 ± 1	−5.3 ± 1.1	−3 ± 0.9	3.3 ± 0.3
	8 weeks	−4.4 ± 2	−1.1 ± 0.4‡	1.6 ± 0.6	1 ± 0.8‡	−1.5 ± 0.6‡	1.9 ± 0.8‡
DC	0 week	−7.1 ± 2.2*	−3.6 ± 0.9	3.3 ± 1.4	−6.9 ± 1.8*	−3.1 ± 1.2	4.4 ± 0.7
	8 weeks	−6.4 ± 2.4*	−3.6 ± 1.3*	2.3 ± 0.5	−1.3 ± 1.4*‡†	−1.7 ± 0.3‡†	1.4 ± 0.7‡

*p < 0.05 compared to MI control, ‡p < 0.05 compared to 0 week baseline within the same group, †p < 0.05 compared to remote region within the same group and timepoint

Table 3 Comparison between regurgitant volume, regurgitant fraction and rating by two-dimensional echocardiography. Data are presented as mean ± standard deviation

Echo rating	n	MRI regurgitant volume (mL)	MRI regurgitant fraction (%)
0	5	2.4 ± 2.5	4.6 ± 3.9
1	12	4.5 ± 2.6	8.7 ± 3.9
2	5	9 ± 4.2	20.3 ± 8.6
3	2	25.3 ± 19.6	35 ± 20.3

attachment node. The circumferential direction was defined as the cross product between the longitudinal and radial vectors. CT attachment node displacements from each region (remote or infarct) were averaged to obtain radial, circumferential, and longitudinal displacement values for each papillary muscle within each animal at each timepoint. Average displacements for each group, region, and timepoint are given in Table 2. A repeated-measures ANOVA model was used to calculate test statistics and p-values for each term in the 3-way full-factorial model, considering the 3-way interaction between timepoint, region, and treatment, as well as each 2-way interaction and main effect. Pairwise differences were extracted and reported here for effects that were statistically significant.

Mitral regurgitation (MR) was rated on a scale from 0 to 3 (none to severe) by two-dimensional echocardiography [10] at both 0- and 8-week time points. Additional quantification of MR was performed using short-axis MRI contours of the LV and RV based on the methods described in previous studies [11–13]. Regurgitant volume was calculated as the difference between the LV and RV stroke volume:

$$Regurgitant\ Volume = SV_{LV} - SV_{RV} \tag{3}$$

and the regurgitant fraction was calculated as:

$$Regurgitant\ Fraction = Regurgitant\ Volume/SV_{LV} \tag{4}$$

where SV units are in mL. A comparison between MRI estimates of regurgitant volume, regurgitant fraction, and MR rating by two-dimensional echocardiography is shown in Table 3.

Average values for regurgitant volume and regurgitant fraction within each treatment group and timepoint are shown in Table 4.

Table 4 Comparison of regurgitant volume and regurgitant fraction between treatment groups. Data are presented as mean ± standard deviation

		0 week	8 weeks
Regurgitant volume (mL)	MI	2.7 ± 1.3	14.8 ± 12.2‡
	DC	4.4 ± 3.9	5.0 ± 3.1*
Regurgitant fraction (%)	MI	7.5 ± 3.8	25.7 ± 13.6‡
	DC	8.9 ± 4.2	7.8 ± 4.2*

*p < 0.05 compared to MI control, ‡p < 0.05 compared to 0 week baseline within the same group

Assessment of myocardial thickness, LV dilation, and function in this cohort of animals has been previously reported [6] and will not be reproduced in this abstract.

Conclusion

While the relationship between the progression of ischemic mitral regurgitation and left ventricular remodeling after myocardial infarction has been investigated in depth, changes in quantifiable metrics of ventricular wall and papillary muscle motion are not as readily available. Previous research has shown that chordae tendineae (CT) have an important role in mitral valve competence as well as left ventricular geometry and function [14]. Due to tethering between the mitral valve and the papillary muscles in the left ventricle via CT, it is plausible that changes in left ventricular wall and papillary muscle motion following myocardial infarction [1] could contribute to the progression of ischemic mitral regurgitation. This study presents a method to quantify changes in papillary muscle motion following posterolateral myocardial infarction.

In this study, quantifiable changes in CT papillary muscle attachment site displacements were assumed to be representative of overall papillary muscle motion. Changes in papillary muscle motion and estimates of mitral regurgitation were compared between a group of animals that received a hydrogel treatment (DC) shown to limit ventricular remodeling after myocardial infarction and a control group (MI). Regurgitant volume and regurgitant fraction calculated from MRI estimates of LV and RV stroke volume (Table 4) were not different between treatment groups at the 0 week baseline timepoint. Both regurgitant volume and regurgitant fraction significantly increased in the untreated MI group from the 0 to 8 weeks timepoint, while no significant change was observed in the treated DC group. Average CT papillary muscle attachment node displacement in the radial, circumferential, and longitudinal directions were compared between treatment groups (MI, DC), region (Remote, Infarct), and timepoint (0 week, 8 weeks) as shown in Table 2. The magnitude of radial displacement was significantly greater in the treated DC group than the untreated MI group regardless of timepoint or region. Despite this systematic difference between groups, it is important to note that the calculated outward radial displacement of the papillary muscle in the infarct region is indicative of an outward bulging of the compromised infarcted tissue region at the 8-weeks timepoint in the untreated MI group. This outward radial displacement calculated from the FE model simulation is consistent with recorded observations of left ventricular dyskinesia from MRI and echocardiography. The magnitude of circumferential displacement significantly decreased between 0- and 8-week time points in the infarct region in both groups, as well as the remote region in the untreated MI group. Circumferential displacement was preserved between 0 and 8 weeks in the remote region of the treated DC group. Longitudinal displacements similarly decreased in the infarct region between 0 and 8 weeks for both treated and untreated groups. Although the magnitude of displacement decreases in all directions in the infarct region for both the MI and DC groups, the positive (outward) motion in the radial direction of the MI group could contribute to the significant increase in regurgitant volume. Additionally, circumferential displacement, which is linked to the twisting motion of the LV, is decreased in both the remote and infarct region of the MI group but is preserved in the remote region of the DC group. The combination of these two factors may serve to highlight the role of changes in papillary muscle motion in the progression of mitral regurgitation. A correlation analysis between measurements of mitral regurgitation and changes in direction and magnitude of displacement may provide some additional insight into the results outlined in this study. Regardless of the underlying mechanism, the results of this study show that regurgitant volume did not increase 8 weeks after myocardial infarction in the group of animals that received the hydrogel injection therapy in contrast to the significant increase in regurgitant volume observed in the untreated animal group.

Acknowledgements This study was supported by National Institutes of Health grants R01 HL063954 (R. Gorman) and U01 HL133359 (J. Wenk).

References

1. M.G.S.J. Sutton, N. Sharpe, Left ventricular remodeling after myocardial infarction. Circulation **101**(25), 2981–2988 (2000). https://doi.org/10.1161/01.cir.101.25.2981
2. F. Bursi et al., Mitral regurgitation after myocardial infarction: a review. Am. J. Med. **119**(2), 103–112 (2006). https://doi.org/10.1016/j.amjmed.2005.08.025
3. G.A. Lamas, Clinical significance of mitral regurgitation after acute myocardial infarction. Survival and ventricular enlargement investigators. Circulation **96**(3), 827–833 (1997)
4. J.F. Wenk et al., First finite element model of the left ventricle with mitral valve: insights into ischemic mitral regurgitation. Ann. Thorac. Surg. **89**(5), 1546–1553 (2010). https://doi.org/10.1016/j.athoracsur.2010.02.036
5. H. Wang et al., Effects of hydrogel injection on borderzone contractility post-myocardial infarction. Biomech. Model. Mechanobiol. **17**(5), 1533–1542 (2018). https://doi.org/10.1007/s10237-018-1039-2
6. C.B. Rodell et al., Injectable shear-thinning hydrogels for minimally invasive delivery to infarcted myocardium to limit left ventricular remodeling. Circ. Cardiovasc. Interv. **9**(10), e004058 (2016). https://doi.org/10.1161/circinterventions.116.004058
7. J.M. Guccione et al., Passive material properties of intact ventricular myocardium determined from a cylindrical model. J. Biomech. Eng. **113**(1), 42 (1991). https://doi.org/10.1115/1.2894084
8. J.M. Guccione, A.D. McCulloch, Mechanics of active contraction in cardiac muscle: part I—constitutive relations for fiber stress that describe deactivation. J. Biomech. Eng. **115**(1), 72 (1993). https://doi.org/10.1115/1.2895473
9. J.M. Guccione et al., Mechanics of active contraction in cardiac muscle: part II—cylindrical models of the systolic left ventricle. J. Biomech. Eng. **115**(1), 82 (1993). https://doi.org/10.1115/1.2895474
10. K.S. Dujardin et al., Grading of mitral regurgitation by quantitative doppler echocardiography. Circulation **96**(10), 3409–3415 (1997). https://doi.org/10.1161/01.cir.96.10.3409
11. W.G. Hundley et al., Magnetic resonance imaging assessment of the severity of mitral regurgitation. Circulation **92**(5), 1151–1158 (1995). https://doi.org/10.1161/01.cir.92.5.1151
12. M.W. Kon, J. Heart Valve Dis. **13**(4), 600–607 (2004)
13. M. Soleimani et al., Moderate mitral regurgitation accelerates left ventricular remodeling after posterolateral myocardial infarction. Ann. Thorac. Surg. **92**(5), 1614–1620 (2011). https://doi.org/10.1016/j.athoracsur.2011.05.117
14. J.F. Obadia et al., Mitral subvalvular apparatus. Circulation **96**(9), 3124–3128 (1997). https://doi.org/10.1161/01.cir.96.9.3124

Characterization of Fiber Alignment and Mechanical Properties of Printed Cellulose Nanofibril Films

Lisa M. Mariani, Gnana Saurya Vankayalapati, John M. Considine, and Kevin T. Turner

Abstract Cellulose nanofibrils (CNFs) are a naturally abundant polymer with exceptional mechanical properties for their low density. Neat CNF materials have been reported with moduli ranging from 4 to 86 GPa, where the variation in moduli results from several preparation parameters, one of which is the fiber alignment. Because of their high aspect ratio (>100), CNFs form an entangled network in the absence of mechanisms for fiber alignment. In this study, the alignment of CNF fibers in films is achieved via control of printing and drying processes used to manufacture neat CNF films from aqueous suspensions containing low volume fractions of CNFs. The alignment of the CNFs is determined both globally and locally within printed CNF thin films and the effect of orientation on mechanical properties is characterized. Polarized light microscopy is used to characterize the orientation of CNFs through the bulk of the material (i.e., over areas >4 mm^2) and shows that propagation of drying fronts can significantly impact alignment. The alignment of CNFs at the surface of the materials is imaged and quantified via atomic force microscopy (AFM). Both topographic and phase imaging, as well as different image processing techniques were evaluated for alignment characterization via AFM.

Keywords Atomic force microscopy · Polarized light microscopy · Nanocellulose · Digital image correlation · Orientation

Introduction

Cellulose nanofibrils (CNFs) are derived from trees, have diameters ~10 nm, and have been shown to have exceptionally high specific stiffness and specific strength as a result of their high aspect ratios (>100), making them an attractive engineering material [1]. Many studies have prepared neat thin films and filaments comprised of CNFs and probed their mechanical properties; however, there is a large variation in tensile moduli and strengths reported for these materials, ranging from 4 to 86 GPa [2–5]. The presence of varying degrees of CNF orientation within the materials is likely one cause of this large disparity of properties reported. Previous studies have demonstrated the orientation of cellulose nanomaterials in both composite and neat materials in a qualitative sense via polarized light microscopy [6–8]. Composites of nanocellulose prepared through printing processes exhibit nanocellulose orientation because of the shear thinning nature of the inks [6]. Whereas, in neat nanocellulose materials prepared in dishes, cellulose nanofibrils align because of surface tension and capillary forces, which arise during drying from the large removal of water volume (>99%) [9].

This study uses polarized light microscopy to assess the global orientation through the bulk in cellulose nanofibril thin films and atomic force microscopy phase imaging to capture and quantify the alignment locally on the surface of each film. The orientation of CNF films made through dish drying, printing at room temperature, and printing at elevated temperature is investigated.

L. M. Mariani (✉) · G. S. Vankayalapati · K. T. Turner
Department of Mechanical Engineering and Applied Mechanics, University of Pennsylvania, Philadelphia, PA, USA
e-mail: lmariani@seas.upenn.edu; saurya7@seas.upenn.edu; kturner@seas.upenn.edu

J. M. Considine
Forest Products Laboratory, USDA Forest Service, Madison, WI, USA
e-mail: jconsidine@fs.fed.us

M. E. Grady (ed.), *Mechanics of Biological Systems and Materials & Micro-and Nanomechanics, Volume 4*,
Conference Proceedings of the Society for Experimental Mechanics Series, https://doi.org/10.1007/978-3-030-30013-5_5

Methods

Thin films comprised of CNFs are prepared via dish drying, printing and drying at room temperature (22 °C), and printing and drying at 70 °C. For dish drying, aqueous suspensions of 1.1 wt % TEMPO pre-treated CNFs are diluted to 0.4 wt % via the addition of deionized water, mechanically stirred, and poured into polystyrene petri dishes. The bubbles are removed via vacuum and then the CNF solutions in the dishes are dried at room temperature in a laminar flow cabinet. For the printed films, 1.1 wt% CNF suspensions are printed on a modified commercial printer (Makerfront). The suspensions are loaded into centrifuge tubes and the bubbles are removed through centrifugation at 3000 rpm for 3 min. The suspensions are then pressured at a pressure of 172.4 kPa through a nozzle of 0.43 mm diameter and deposited on glass silanized build plates. The specimens are dried at either 22 °C or 70 °C. Printed films used for orientation characterization consist of 2 layers.

Orientation is assessed qualitatively through polarized light microscopy (PLM) and quantitatively through atomic force microscopy (AFM) tapping mode phase imaging. The polarized light microscope (Leica DMRX) is equipped with a 5× objective and a full wave plate. The specimens are imaged on a glass slide between crossed polarizers and the light passes through the wave plate and then through the birefringent specimen. Images are collected with the crossed polarizers perpendicular to the edge of the rectangular specimens and with the specimens rotated 45% relative to the cross polarizers. The specimens are then probed using tapping mode atomic force microscopy (Bruker Dimension Icon) with tips <10 nm in diameter (Arrow NCR). Images are collected with a minimum scan resolution of 5 nm/pixel. As the tip scans in the valleys between CNFs it feels an attractive force because the fiber diameter is on the same order as the tip diameter. This attractive force results in a phase lag between the cantilever drive and response signal. Thus, the phase image gives information about the orientation of the individual fibers because of the phase differences detected at the edge of each CNF. Image processing of the phase images is then completed in Matlab to quantify the orientation.

Results and Discussion

CNF films are cut in rectangular films and mounted onto glass slides for PLM and AFM analysis. Figure 1a shows a representative image from PLM, where there is a CNF film (orange) on a glass slide (purple). The direction of the wave plate orientation is given by the blue arrow. An un-oriented film would appear as the same color as the glass slide. The orange color in the CNF thin film indicates that the CNF orientation is more closely aligned with the direction of the red arrow. PLM reveals that all three film preparation methods (dish drying, printing at 22 °C, printing at 70 °C) show some alignment of CNFs. The films with higher degrees of alignment have a greater intensity of light that passes through the polarizers (ex. brighter orange indicates more orientation than dark orange). The highest intensity of orientation in PLM is found in CNF films printed at dried at 70 °C; whereas the light intensity of the dish dried and room temperature printed films are similar. Additionally, for all specimens the orientation is found to be primarily parallel to the contact line formed during drying.

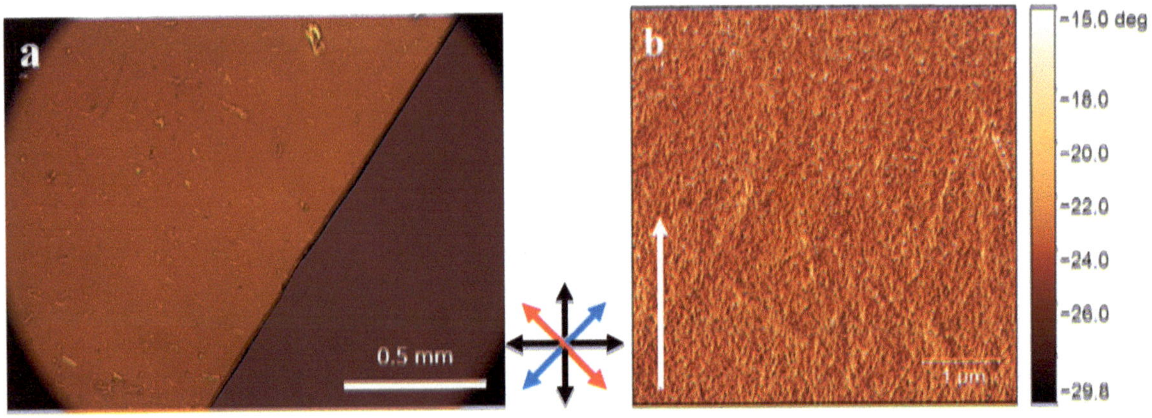

Fig. 1 (**a**) Polarized light microscopy image of orientation within a cellulose nanofibril thin film. The direction of crossed polarizers is indicated by the black arrows and the red and blue arrows indicate the direction of fiber orientation. (**b**) atomic force microscope phase image for an oriented cellulose nanofibril thin film [10]. The image size is 5 × 5 μm with a resolution of 1024 samples/line

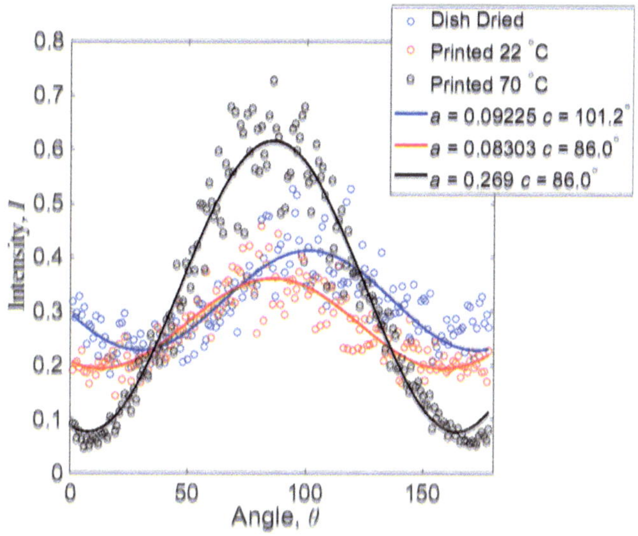

Fig. 2 Representative intensity data and cosine fits for thin films prepared via dish drying, printing at 22 °C, and printing at 70 °C. The circles are the data and the lines are the cosine fits. The relevant fit parameters are given in the legend

The orientation is further investigated by AFM imaging. Figure 1b shows a representative AFM phase image of an oriented CNF film, where the orientation is in the direction of the white arrow given in the figure.

The AFM phase images are loaded imported into Matlab for image processing. A radial sums of the intensity values of the 2D Fourier transform of the grey-scale AFM phase images is used to examine the intensity of the orientation angles of fibers for the CNF thin films. The intensity of orientation as a function of the angle is fit with a cosine function, given in eq. 1, where *a, b, c,* and *d* are fitting parameters.

$$I = \mathrm{a} \cdot \cos\left(b\left(\theta - c\right)\right) + d \tag{1}$$

The intensity of the orientation is given by *a*, and the orientation angle is given by *c*. Figure 2 shows representative plots of the intensities as function of orientation angle and the corresponding cosine fits for each CNF thin film preparation condition. The orientation is highest for films printed and cured at 70 °C. The films printed at room temperature and those dried in petri dishes at room temperature have similar fits, which both correspond to weak orientation. This is not surprising, as the surface tension forces causing orientation are weaker at room temperature than at elevated temperatures.

Conclusion

This research links the orientation of CNF films characterized via polarized light microscopy to the orientation seen in atomic force microscopy phase imaging. Through AFM phase imaging, the degree and intensity of alignment in the CNF materials is quantified and compared in thin films prepared via various techniques. The CNF thin films printed at 70 °C have a higher degree of orientation than those prepared via dish drying, as a result of the drying mechanics. Additionally, those films printed at an elevated temperature have more orientation than those printed at room temperature. The ability to quantify the orientation in CNF thin films allows the variation in mechanical properties to be compared by the respective orientation of specimens tested.

Acknowledgements The authors would like to acknowledge support from the U.S. Endowment for Forestry & Communities, Inc. as part of the P3Nano program. This work was performed in part at the University of Pennsylvania's Singh Center for Nanotechnology, an NNCI member supported by NSF Grant ECCS-1542153. The authors thank Dr. Arjun Yodh at the University of Pennsylvania for access to polarized light microscopy. Dr. Craig Clemons at the USDA Forest Products Laboratory also provided helpful insight on interpretation of polarized light microscopy results.

References

1. A. Isogai, Wood nanocelluloses: fundamentals and applications as new bio-based nanomaterials. J. Wood Sci. **59**(6), 449–459 (2013)
2. H. Sehaqui, N. Ezekiel Mushi, S. Morimune, M. Salajkova, T. Nishino, L.A. Berglund, Cellulose nanofiber orientation in nanopaper and nanocomposites by cold drawing. ACS Appl. Mater. Interfaces **4**(2), 1043–1049 (2012)
3. C. Baez, J. Considine, R. Rowlands, Influence of drying restraint on physical and mechanical properties of nanofibrillated cellulose films. Cellulose **21**(1), 347–356 (2014)
4. M. Henriksson, L.A. Berglund, P. Isaksson, T. Lindstrom, T. Nishino, Cellulose nanopaper structures of high toughness. Biomacromolecules **9**(6), 1579–1585 (2008)
5. K.M.O. Hakansson, I.C. Henriksson, C.d.l.P. Vazquez, V. Kuzmenko, K. Markstedt, P. Enoksson, P. Gatenholm, Solidification of 3D printed nanofibril hydrogels into functional 3D cellulose structures. Adv. Mater. Technol. **1**, 1600096 (2016)
6. G. Siqueira, D. Kokkinis, R. Libanori, M.K. Hausmann, A.S. Gladman, A. Neels, P. Tingaut, T. Zimmermann, J.A. Lewis, A.R. Studart, Cellulose nanocrystal inks for 3D printing of textured cellular architectures. Adv. Funct. Mater. **27**(12), 1604619 (2017)
7. K.M.O. Håkansson, A.B. Fall, F. Lundell, S. Yu, C. Krywka, S.V. Roth, G. Santoro, M. Kvick, L. Prahl Wittberg, L. Wågberg, L.D. Söderberg, Hydrodynamic alignment and assembly of nanofibrils resulting in strong cellulose filaments. Nat. Commun. **5**, 4018 (2014)
8. F. Ansari, C. Brouzet, S.V. Roth, L. Wågberg, P. Chen, K. Gowda.V, N.A. Kotov, N. Mittal, L.D. Söderberg, P.T. Larsson, F. Lundell, Multiscale control of nanocellulose assembly: transferring remarkable nanoscale fibril mechanics to macroscale fibers. ACS Nano **12**(7), 6378–6388 (2018)
9. A. Skogberg, A.J. Mäki, M. Mettänen, P. Lahtinen, P. Kallio, Cellulose nanofiber alignment using evaporation-induced droplet-casting, and cell alignment on aligned nanocellulose surfaces. Biomacromolecules **18**(12), 3936–3953 (2017)
10. L.M. Mariani, W.R. Johnson, J.M. Considine, K.T. Turner, Printing and mechanical characterization of cellulose nanofibril materials. Cellulose **26**(4), 2639–2651 (2019)

Vibrational Analysis of Biopolymer-Based Hydrogels Using 3D-Printed Test Structures for Applications in Bioprinting

S. Schwarz, B. Hartmann, R. Moerl, S. Sudhop, H. Clausen-Schaumann, and D. Rixen

Abstract The mechanical properties of hydrogels suitable for applications in the field of bioprinting, which tries to develop three-dimensional tissue equivalents, are crucial for the proper fulfilment of their functions in the human body. This aspect is especially important regarding types of tissues which have to withstand applied mechanical forces. Due to their high water content similar to the human body and their tunable mechanical properties, hydrogels based on biopolymers are ideally suited for such applications. In this work, the first results of a novel method for the indirect measurement of the mechanical properties of hydrogels using laser-Doppler vibrometry and 3D-printed test structures are presented. Thanks to the experimental design hydrogels can be cast directly over such beam-like test structures without any leakage. First results show that the resonance frequencies of the beam structure are modulated by the material properties of the different hydrogels placed on it, enabling future applications and further experiments. For comparing the measurement data with the mechanical properties of the samples used, indentation-based measurements have been carried out. This approach can be integrated into existing bioprinting workflows and enables the non-destructive monitoring of biopolymer-based hydrogels in their mechanical properties.

Keywords Laser-Doppler vibrometry · Hydrogels · Vibrational analysis · Bioprinting · Additive manufacturing

Introduction

Due to their high water content similar to the human body and their tunable mechanical properties, hydrogels based on biopolymers are ideally suited for applications in the field of bioprinting. These hydrogels act as bioorganic support material, where living cells and bioactive growth factors are embedded to generate three-dimensional tissue equivalents, which can be used as replacements for damaged tissues in the regenerative medicine or as tests systems for toxicity assay or drug screening in the pharmaceutical industry [1–3]. The mechanical properties of the used hydrogel are crucial for the functionality of the tissue equivalent: The hydrogel defines the microenvironment of the embedded cells, and its viscoelastic properties have a direct influence on cell fate and therefore correct morphogenesis of the artificial tissue [4, 5]. Furthermore, the mechanical stiffness of the completed tissue construct is a crucial parameter for its biomechanical function in the human body since it must be capable of withstanding applied forces. Additionally, nature often solves problems by introducing a mechanical, stiffness dependent anisotropy. This can be observed in hard-soft transitions like those that can be found in areas where tendons are inserting into the bone [6]. However, especially these areas prevent the whole system from failure due to high forces applied. For measuring the mechanical properties of hydrogels directly during the bioprinting process, a technique that can be directly integrated into this workflow would be desirable.

S. Schwarz (✉)
Center for Applied Tissue Engineering and Regenerative Medicine–CANTER, Munich University of Applied Sciences, Munich, Germany

Department of Mechanical Engineering, Technical University of Munich, Garching, Germany
e-mail: Sascha.schwarz@hm.edu

B. Hartmann · S. Sudhop · H. Clausen-Schaumann
Center for Applied Tissue Engineering and Regenerative Medicine–CANTER, Munich University of Applied Sciences, Munich, Germany, Germany

R. Moerl
Polytec GmbH, Waldbronn, Germany

D. Rixen
Department of Mechanical Engineering, Technical University of Munich, Garching, Germany

M. E. Grady (ed.), *Mechanics of Biological Systems and Materials & Micro-and Nanomechanics, Volume 4*,
Conference Proceedings of the Society for Experimental Mechanics Series, https://doi.org/10.1007/978-3-030-30013-5_6

Mechanical characterization of hydrogels and artificial tissues constructs is usually done by contact-based methods like indentation and dynamical mechanical analysis [7]. This frequently leads to damage of the sample due to the applied force. Additionally, contact with the measuring tool (indenter, device) can be a source for biological contamination, so that the tissue construct cannot be used any further. Therefore, a simple, contamination-free and non-destructive material testing is a significant advantage, especially in tissue engineering applications, where expensive biomaterials and human cells are used.

As a laser-based, non-destructive and high-resolution measurement technique Laser-Doppler Vibrometry (LDV) is already established in the vibrational characterization in a broad variety of applications [8, 9]. In the area of biomedical engineering LDV can provide new insights into the function of biological systems [10–12]. The direct measurement of vibrations in optical transparent hydrogels is so far not possible, due to the small amount of laser intensity being reflected. To overcome this circumstance reflective coatings, stickers or encapsulated reflective particles are often used [13, 14]. When it comes to soft materials, these methods carry a particular risk of altering the mechanical properties due to additional mass or stiffness added to the system.

In this paper the first results of a new and indirect method for the measurement of mechanical properties of hydrogels are presented. With the help of additive manufacturing, a hydrogel sample holder with an integrated beam, which is fixed at both ends and acts as a "cantilever" is constructed. Gelatin hydrogels are cast directly into this sample holder. After gel solidification, this sample holder is positioned on top of a piezo stack for excitation, and the vibration of the integrated beam is measured using a LDV. With this setup, hydrogels of various Gelatin concentration and therefore with different stiffness were analyzed by LDV. We were able to show that the vibration of the beam and its change due to the material properties of the hydrogel samples on top can be measured using a LDV, without the need to use additives to increase the reflectivity of the hydrogels. To correlate the mechanical properties of the prepared hydrogel samples with the experimental LDV data, indentation-type atomic force microscopy (IT-AFM) experiments were used for determining the Young's modulus of the different samples.

Materials and Methods

Printing of Sample Holder

Sample holder with the integrated beam geometry (Fig. 1d) was manufactured using a Form 2 stereolithographic 3D-printer (Formlabs Inc., Somerville, MA, USA), clear resin version 4 and a layer height of 0.25 mm. After printing, the model was washed using the Form wash station (Formlabs Inc., Somerville, MA, USA) and cured for 30 min at 60 °C using the Form curing station (Formlabs Inc., Somerville, MA, USA). The thread necessary to screw the sample holder directly onto the piezo stack was cut directly into the resin material. For the dimensions of the beam, a width and thickness of 1 mm and a length of 12 mm were chosen, as these dimensions could be printed without deformation of the beam, due to internal stresses, which may occur during the curing process. In order to avoid possible clogging of the thin gaps lateral of the beam during the printing process, the smallest possible dimensions for these gaps were determined and finally set to 0.4 mm.

Preparation of Gelatin Samples

To prepare for example 0.1 g/ml hydrogel stock solution, 1 g of Gelatin powder (4308.3, 160 bloom, Carl Roth GmbH, Karlsruhe, Germany) were gently mixed with 10 ml deionized water and placed for 20 min at a magnetic stirrer tempered at 60 °C. In order to minimize the evaporation effects during sample preparation, the beaker glasses were covered with Parafilm® (P7793, Sigma-Aldrich, Taufkirchen, Germany). After the gelatin was completely melted, the samples were stored in an incubator (37 °C, 80% relative humidity) until they were used for the experiments.

Preparation of Sealing Material

The sealing material was made by mixing 3.3 g of Pluronic® F-127 powder (P2443-250G, Sigma-Aldrich, Taufkirchen, Germany) with 10 ml of deionized water manually using a spatula. After the mixture adapted an opaque and greasy-like consistency, it was stored overnight in the fridge at 4 °C. At this concentration, Pluronic® F-127 solidifies as a thermoresponsive material around 37 °C and liquefies at 4 °C. After 12 h at 4 °C, the liquid solution can be poured directly into the syringe and stored again at 4 °C until the material is needed for experiments.

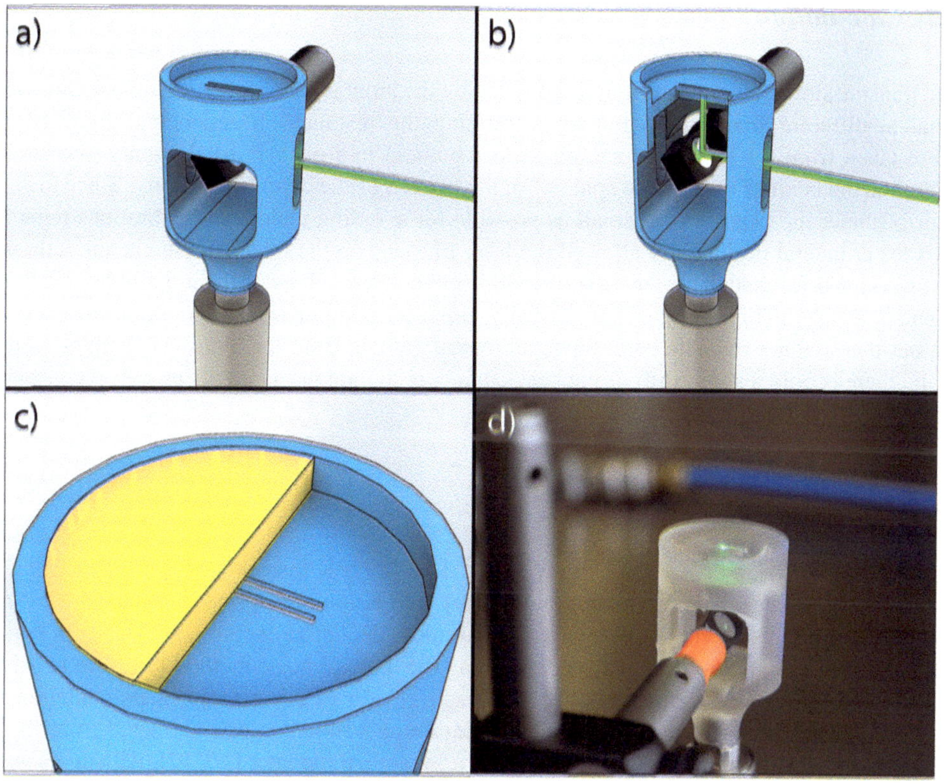

Fig. 1 (**a**) The 3D-printed sample holder (blue) was manufactured using a stereolithographic 3D-printer. The setup is directly screwed onto the piezo stack. (**b**) The mirror can be positioned in the cavity of the sample holder without risking coupling of the mechanical vibrations. The laser beam (green) can be deflected on the back of the beam via the mirror through lateral openings. (**c**) Sectional view of the sample (orange) and beam. The Gelatin (orange) can be pipetted precisely into the cylindrical cavity area of the sample holder. (**d**) Complete, ready-to-use experimental setup

Experimental Setup

Experiments were done using an infrared scanning LDV (PSV-500 Xtra, Polytec GmbH, Waldbronn, Germany) and a piezo stack (P-844.10, Physik Instrumente, Karlsruhe, Germany) for the excitation. As an excitation signal, a frequency sweep from 0 to 25,000 Hz (periodic chirp, amplitude 2 V) was applied using the in-build signal generator. The piezo was fixed using double-sided adhesive tape. The silver mirror (Thorlabs GmbH, Dachau, Germany) was positioned and finally fixed using standard optical post assembly parts (Thorlabs GmbH, Dachau, Germany) onto a solid aluminium breadboard (Thorlabs GmbH, Dachau, Germany). For vibrational decoupling of the setup, this plate was placed onto a 8 mm thick Sorbothane dampening sheet (Thorlabs GmbH, Dachau, Germany). For sealing the two gaps lateral of the beam (Fig. 1c), the liquid sealing material was manually applied using a 10 ml syringe and a standard nozzle with a diameter of 0.33 mm (F560014-1_4, Vieweg GmbH, Kranzberg, Germany) stored at 4 °C. Shortly after the liquid material had been applied into the gaps, it solidified and closed the gaps without significantly affecting the vibrations of the beam. This is mainly because Pluronic® F-127 does not form a polymer network such as biopolymer-based hydrogels, but forms micellar structures due to its amphiphilic molecules, which can be easily shifted towards each other. Afterwards, the Gelatin hydrogel was directly cast into the cylindrical geometry located at the top of the sample holder (Fig. 1c) and placed for 6 min into the fridge at 4 °C for solidification. Afterwards, the sample was directly screwed onto the piezo stack and carefully tightened. Before starting the measurement, there was a resting time of 8 min so the sample can adapt to room temperature (23.8 °C) and avoid thermal equilibration processes which may influence the measurement. During this period, the mirror was positioned for deflecting the laser beam of the LDV from the horizontal 90 degrees upwards onto the middle of the back of the beam (Fig. 1b and c). After the measurement was completed, the sample holder was cleaned using warm water (50 °C) for melting the Gelatin away and dried using pressurized air.

Transmissibility Calculation

For calculating the transmissibility, the output signal is divided by the input signal for getting information about amplification of the input signal at different frequencies. For characterizing the resonance frequencies of the beam structure, first measurements were taken from the top of the sample holder without hydrogel. In total twenty measurement points were placed circularly around the beam (Fig. 2, green spheres) at the bottom of the cylindrical cavity. Since the area spared out for positioning the mirror under the beam was as small as possible for avoiding undesirable vibrations regarding the complete sample holder, this experimental design only allowed the laser to focus on the middle of the beam without much variation. Therefore, as an input signal for calculating the transmissibility the mean values of the amplitude velocity from the twenty measurements points was taken (Fig. 2, blue line). Usually, input and output signal should be measured within the same experimental run, but this was not possible here. Since all measured data from the hydrogel samples were divided through the same set of data, there is only a slight shift of the peaks, but overall data the same offset.

IT-AFM Testing

IT-AFM measurements were carried out using a NanoWizard® I AFM (JPK Instruments, Berlin, Germany). Silicon nitride cantilevers (MLCT Cantilever E, Bruker, Mannheim, Germany) with a nominal spring constant of 0.1 N/m and a nominal tip radius of 20 nm were used. The vertical tip velocity during the indentation experiments was kept constant at 12 μm/s. The spring constant of the used cantilever was determined using the thermal noise method [15] The IT-AFM measurements were preformed in phosphate buffered saline solution. On each gelatin gel, five force maps containing 10 × 10 force-indentation curves and an area of 5 × 5 μm^2 were taken. The Young's modulus was extracted from the approach part of the force-indentation curves using a modified Hertz model for a pyramidal indenter [16, 17]. All curves were analyzed from the contact point up to a maximum indentation depth of 1 μm, using the JPK Data Processing software (Version 6.1.42, JPK Instruments AG, Berlin, Germany). The Young's modulus results of each force map were summarized in a stiffness distribution (histogram). To locate the maxima of these histograms, a Gaussian function was fitted to each histogram using the Igor software (Version 6.3.4.0, WaveMetrics, Oregon, USA).

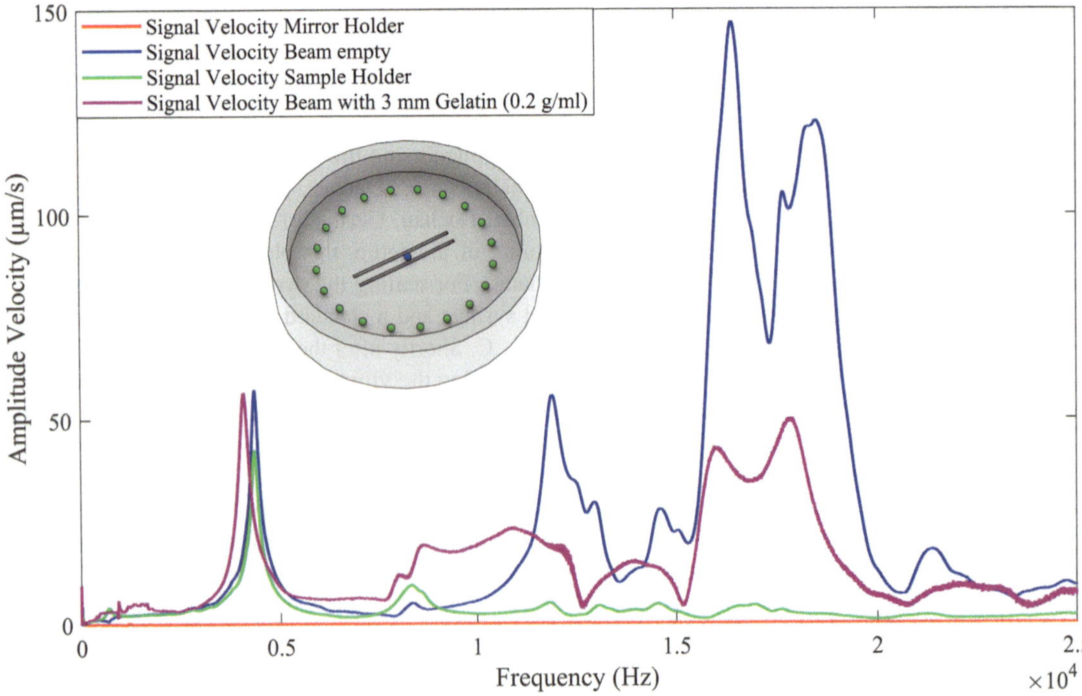

Fig. 2 Vibrational analysis of the different components of the experimental setup. The green dots are representing the area inside of the cylindrical cavity where the measurements have been done. Data shown in the blue line was measured at the center of beam structure (blue dot). The signal from the purple line was taken from the backside of the beam, covered with 3 mm of Gelatin (concentration: 0.2 g/ml). The red line at the bottom shows the vibrations of the mirror holder

Results

First the sample holder was characterized in its behaviour placing 20 measurement points (Fig. 1, green dots) at the bottom of the cylindrical cavity and one measurement point (Fig. 1, blue dot) onto the middle of the beam. Resulting signal data represented in blue and green line in Fig. 1 is clearly showing a difference. Considering measurement data of the beam, dynamics that do not occur in the rest of the sample holder can be identified. When a hydrogel is cast upon the beam a clear change in the systems response can be observed. The peak at around 16.4 kHz of the beam signal taken from the middle (Fig. 2, blue line) can be clearly defined, and complete scanning data taken from the beam shows are nicely shaped bending mode with a maximal amplitude of about 116.6 μm/s regarding the velocity and around 1.42 nm. The blue line in Fig. 2 represents the mean value of the data points taken around the beam structure, illustrated by the blue spheres in the image, and shows a clear peak around 4.3 kHz which represents the first resonance of the complete sample holder. When laden with a hydrogel (Fig. 2, purple line) there is a definite change in signal and also a slight shift of this first resonance frequency. This first dynamic of the system is not of interest to us, as it does not limit to the beam and thus no direct interaction with the respective test hydrogel. The measurement data shows above all a good signal quality, although the measurement was carried out on a transparent material. For evaluating the vibrational decoupling of the mirror and the piezo as the excitation source, a measurement directly on the holder of the mirror was done. The result (Fig. 2, red line) is clearly showing no significant signal during the measurement, and so there is a complete decoupling and no interfering dynamics from the holder, which could negatively influence the relevant measurements.

The transmissibility data (Fig. 3) of the empty beam shows three clear main peaks between 10 and 20 kHz, representing the resonance frequencies of this geometry. Compared to the data of the different Gelatin concentrations there is an evident change in the signal, but no clear change among the different hydrogel samples. The first small peak at 4 kHz is not of interest, since this represents the first resonance frequency of the whole sample holder and not only the beam structure. This statement is only possible using a scanning LDV, because here the individual mode shapes can be analyzed very precisely and thus assigned to the respective geometric areas. The slight shift observable when comparing the data of the empty beam and the data of the samples is a result of dividing the signal of the samples with the mean signal without the samples. This is not optimal, but in this first experimental design, this was the most effective way of calculating the transmissibility. Compared with the other resonance frequencies only the second at 16.4 kHz was good enough for further data analysis. The peak situated towards higher frequencies shows a similar behavior regarding the decrease of peak heights with increasing polymer concentrations, but due to the bimodal distribution data analysis would have been more prone to errors.

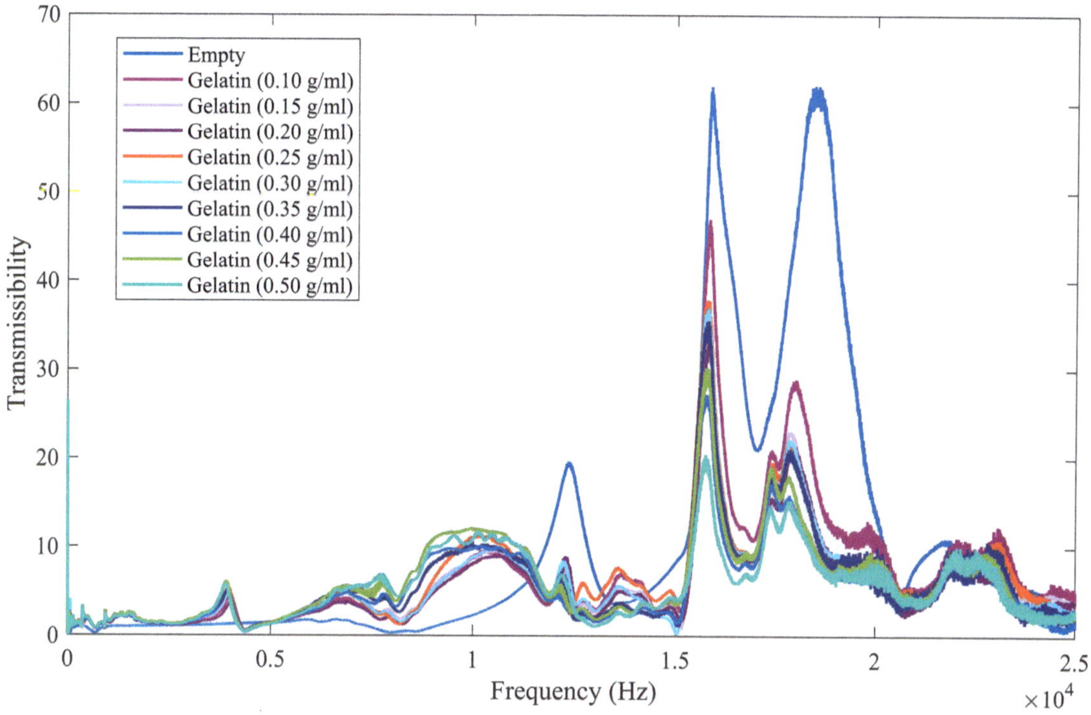

Fig. 3 Comparison of the transmissibility (output/input) from different Gelatin samples and of the empty beam

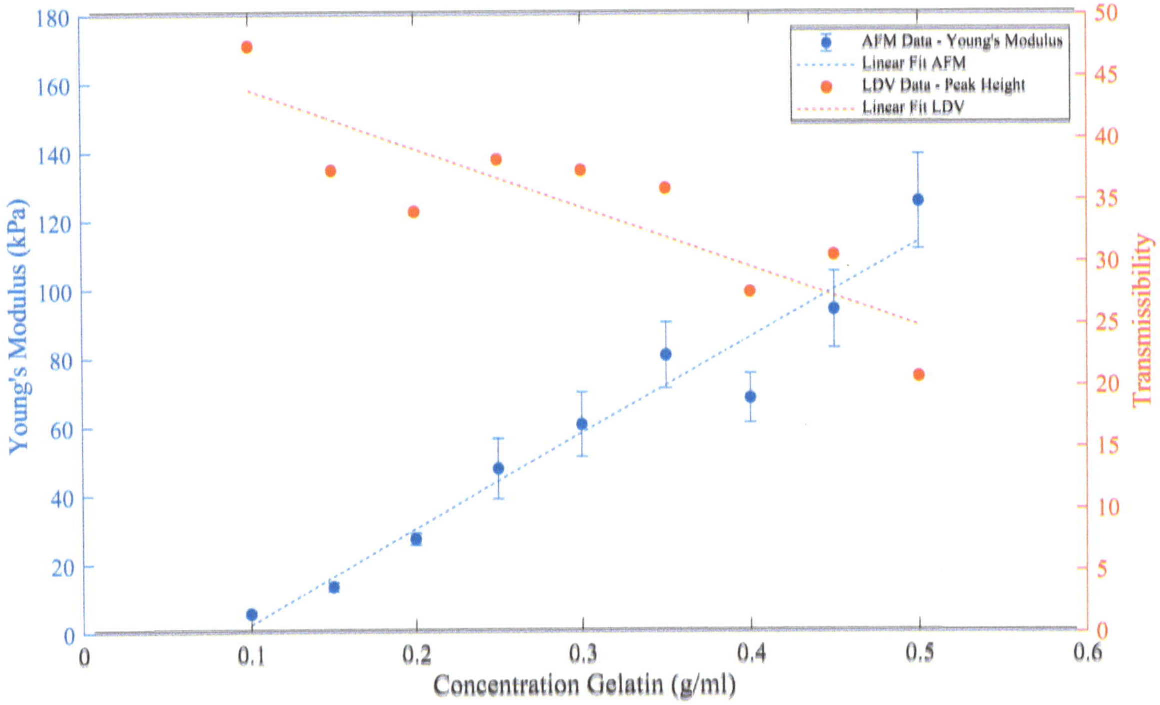

Fig. 4 Comparison of the mechanical properties measured using IT-AFM and the peak height of the second resonance frequency at 16.4 kHz from Fig. 3

For further data analysis and to finally correlate the changing mechanical properties of the different Gelatin samples, the peak height of the second resonance frequency at 16.4 kHz and Young's modulus measured with the IT-AFM were compared (Fig. 4). With higher polymer concentrations, also the stiffness of the Gelatin increases linearly. In contrast to this, the peak height behaves opposite and decreases with increasing stiffness. Since the peak height represents the amplitude of vibration regarding the velocity this makes sense. The stiffer the hydrogel becomes with increasing concentration, the more it can counteract the penetration movement of the beam. In addition to this parameter, the full width at half maximum and frequency shift of the peaks were also considered, but no significant change of those parameters could be observed.

Conclusion

Results are showing the feasibility of the here presented approach for characterizing the mechanical properties of hydrogels using a small, 3D-printed beam (1 × 1 × 12 mm). Changes in the material properties of different hydrogel samples can be measured indirectly through the vibrational analysis of a beam-like structure using LDV. Additionally, it was successfully shown that objects 3D-printed out of transparent materials could be measured directly. Through the use of a fugitive material, also the measurement of fragile, in the beginning liquid hydrogels is possible without any leakage. This enables for example the characterization of self-polymerizing kinetics many biopolymers possess.

Experimental results are showing beside the decrease of peak heights various changes when comparing the different polymer concentrations among each other. A closer look at this testing system and its optimization might allow further statements to be made about the mechanical properties of tested materials. If one considers the low amplitude of about 1.4 nm appearing at 16.4 kHz, it is questionable to what extent a meaningful and reproducible interaction of the beam and hydrogel is taking place. It makes sense that the penetration depth of the beam into the hydrogel, and therefore the peak height decreases with increasing stiffness, but one would have expected a shift of the resonance frequencies and a change of the modal damping or the half-value width of the relevant peaks. Given the test results, the test design should be revised again in order to correct the disadvantages mentioned. With high probability, the beam used here is too stiff compared to the test hydrogels for measuring all of the changes. A thinner, more fragile and therefore sensitive beam structure could bring significant improvements here. A one-sided fixed beam, which would result in a cantilever, could achieve a higher amplitude and thus possibly a greater interaction with the material of interest. This could lead to a clearer response regarding changing mechanical properties.

Regarding future applications, this approach can be easily integrated into different processes where the material properties of hydrogels have to be monitored for ensuring a proper product, like for example Bioprinting. By directly printing hydrogel strands onto multiple beam structures and measuring the resulting modulation of their resonance frequencies using a scanning LDV, the mechanical properties could be analyzed directly during the printing process. This would enable the production of three-dimensional structures possessing defined mechanical properties, necessary for enabling proper tissue growth and therefore the production of functional tissue equivalents.

Acknowledgements The authors acknowledge financial support through the research focus *"Herstellung und biophysikalische Charakterisierung drei-dimensionaler Gewebe – CANTER"* of the Bavarian State Ministry for Science and Education and the financial support through the *"BayWISS – Ressourceneffizienz und Werkstoffe"* program.

References

1. J. Malda, J. Visser, F.P. Melchels, T. Jüngst, W.E. Hennink, W.J.A. Dhert, J. Groll, D.W. Hutmacher, 25th anniversary article: engineering hydrogels for biofabrication. Adv. Mater. **25**, 5011–5028 (2013). https://doi.org/10.1002/adma.201302042
2. J.K. Carrow, P. Kerativitayanan, M.K. Jaiswal, G. Lokhande, A.K. Gaharwar, Polymers for bioprinting, in *Essentials 3D biofabrication translation*, (Elsevier, Amsterdam, 2015), pp. 229–248. https://doi.org/10.1016/B978-0-12-800972-7.00013-X
3. A.S. Hoffman, Hydrogels for biomedical applications. Adv. Drug Deliv. Rev. **64**, 18–23 (2012). https://doi.org/10.1016/j.addr.2012.09.010
4. A.J. Engler, S. Sen, H.L. Sweeney, D.E. Discher, Matrix elasticity directs stem cell lineage specification. Cell **126**, 677–689 (2006). https://doi.org/10.1016/j.cell.2006.06.044
5. J. Swift, I.L. Ivanovska, A. Buxboim, T. Harada, P.C.D.P. Dingal, J. Pinter, J.D. Pajerowski, K.R. Spinler, J.W. Shin, M. Tewari, F. Rehfeldt, D.W. Speicher, D.E. Discher, Nuclear lamin-A scales with tissue stiffness and enhances matrix-directed differentiation. Science **341** (2013). https://doi.org/10.1126/science.1240104
6. L. Rossetti, L.A. Kuntz, E. Kunold, J. Schock, H. Grabmayr, S.A. Sieber, R. Burgkart, A.R. Bausch, The microstructure and micromechanics of the tendon–bone insertion. Nat. Mater. **16**(6), 664–670 (2017). https://doi.org/10.1038/NMAT4863
7. M.L. Oyen, Mechanical characterisation of hydrogel materials. Int. Mater. Rev. **59**, 44–59 (2014). https://doi.org/10.1179/1743280413Y.0000000022
8. S.J. Rothberg, M.S. Allen, P. Castellini, D. Di Maio, J.J.J. Dirckx, D.J. Ewins, B.J. Halkon, P. Muyshondt, N. Paone, T. Ryan, H. Steger, E.P. Tomasini, S. Vanlanduit, J.F. Vignola, An international review of laser Doppler vibrometry: Making light work of vibration measurement. Opt. Lasers Eng. **99**, 11–22 (2016). https://doi.org/10.1016/j.optlaseng.2016.10.023
9. P. Castellini, M. Martarelli, E.P. Tomasini, Laser doppler vibrometry: development of advanced solutions answering to technology's needs. Mech. Syst. Signal Process. **20**, 1265–1285 (2006). https://doi.org/10.1016/j.ymssp.2005.11.015
10. J.J. Rosowski, R.P. Mehta, S.P. Merchant, Diagnostic utility of laser-doppler vibrometry in conductive hearing loss with normal tympanic membrane. Otol. Neurotol. **24**, 165–175 (2003). https://doi.org/10.1158/0008-5472.CAN-10-4002.BONE
11. T. Schuurman, D.J. Rixen, C.A. Swenne, J.W. Hinnen, Feasibility of laser doppler vibrometry as potential diagnostic tool for patients with abdominal aortic aneurysms. J. Biomech. **46**, 1113–1120 (2013). https://doi.org/10.1016/j.jbiomech.2013.01.013
12. N.E. Conza, D.J. Rixen, S. Plomp, Vibration testing of a fresh-frozen human pelvis: The role of the pelvic ligaments. J. Biomech. **40**, 1599–1605 (2007). https://doi.org/10.1016/j.jbiomech.2006.07.001
13. F.C. Meral, T.J. Royston, R. Magin, Fractional calculus in viscoelasticity: An experimental study. Commun. Nonlinear Sci. Numer. Simul. **15**, 939–945 (2010). https://doi.org/10.1016/j.cnsns.2009.05.004
14. G. Xu, C. Wang, T. Feng, D.E. Oliver, X. Wang, Non-contact photoacoustic tomography with a laser Doppler vibrometer", Proc. SPIE 8943, Photons Plus Ultrasound: Imaging and Sensing 2014, 894332 (3 March 2014); https://doi.org/10.1117/12.2040290
15. M. Jaschke, H.J. Butt, Calculation of thermal noise in atomic force microscopy. Nanotechnology **6**, 1–7 (1995). https://doi.org/10.1088/0957-4484/6/1/001
16. I.N. Sneddon, The relation between load and penetration in the axisymmetric boussinesq problem for a punch of arbitrary profile. Int. J. Eng. Sci. **3**, 47–57 (1965). https://doi.org/10.1016/0020-7225(65)90019-4
17. D.C. Lin, E.K. Dimitriadis, F. Horkay, Robust strategies for automated AFM force curve analysis—I. Non-adhesive indentation of soft, inhomogeneous materials. J. Biomech. Eng. **129**, 430–440 (2006)

High Amplitude Torsional Shear of Porcine Thoracic Aorta

Akshay Rao, Manoj Myneni, C. C. Benjamin, and K. R. Rajagopal

Abstract Studying the mechanics of aortic tissue is a crucial component in understanding its behavior under healthy as well as diseased conditions. Wall shear stress and circumferential stress have been largely accepted as significant factors in arterial growth and remodeling as a response to changes in flow and pressure. But, experimental studies on aortic tissues have largely focused on uniaxial and biaxial tests which are more suited for investigating circumferential stress. This may be explained by the inherent convenience of gripping tissue and then applying deformations in the uniaxial and biaxial modes. As such, the behavior of aortic tissue under shear has been left relatively unstudied. We propose to study the response of porcine aortic wall tissue under cyclic constant torsional shear strain rate for high amplitude (50%) and at different shear strain rates (4%/s and 40%/s). Three to four 12.5 mm diameter samples are obtained from the descending porcine aorta. Initial results clearly indicate a non linear response for the moment as a function of the angle of twist while many popular models predict a linear response for the arterial wall even under large shear strain.

Keywords Shear · Rate-dependence · Torsion · Thoracic aorta · Rheometer

Introduction

In this study we seek to characterize the response of porcine aortic tissue subjected to large amplitude shear. Wall shear stress and circumferential stress are significant contributing factors to arterial growth and remodeling. While the mechanics of these aortic wall has been studied at length these studies have focused on uniaxial and biaxial tests. These tests study circumferential stresses and have become a standard test for biological materials. Test on the aorta's response to a shearing load has been largely unexplored.

The aorta can undergo shearing deformation in both physiological and non physiological situations [2]. Injury from physical impact can cause large shear deformation. In addition normal functioning of the aortas can cause shearing. The pulsatile flow of the blood especially at regions of branching of capillaries. This is in addition to shearing caused by hemodynamic shear stress from blood flow on the arterial wall.

Earlier studies on shear of aortas have largely focused on inflation deflation and torsion by twisting [6, 7] and [10]. While more recent studies have used shear testing devices capable of producing shear simultaneously in two orthogonal directions and measuring the forces in three axes like one on human adipose tissue [8] where shear was applied sinusoidally in orthogonal directions for all six modes of shear. Others have tested human myocardium specimen [9], rat myocardium [3] and pig myocardium [4] using a triaxial tester. Avazmohammadi et al. [1] presented a general approach to study the mechanical behavior of soft materials by applying tensile, compressive or shear loads on a cuboidal myocardium specimen.

Here we aim to lay the ground work for future torsional shear testing using a rheometer on the porcine arterial wall. Porcine tissue was chosen wall as a precursor to possible testing on human tissue due to their anatomical similarities. As such we are investigating the response to shearing at different amplitudes and rates. The first being constant shear rate tests at different amplitudes varying from 10% shear strain upto 50% shear strain in steps of 10%. In addition we are applying two shear rates at 4%/s and 40%/s. Samples are tested from both the lower and upper descending aorta.

A. Rao · M. Myneni · C. C. Benjamin (✉) · K. R. Rajagopal
Department of Mechanical Engineering, Texas A&M University, College Station, TX, USA
e-mail: akshayrao@tamu.edu; manojmyneni@tamu.edu; ccbenjamin@tamu.edu; krajagopal@tamu.edu

M. E. Grady (ed.), *Mechanics of Biological Systems and Materials & Micro-and Nanomechanics, Volume 4*,
Conference Proceedings of the Society for Experimental Mechanics Series, https://doi.org/10.1007/978-3-030-30013-5_7

Methods

Specimen Harvest

Heart and aorta are collected from freshly killed hogs from the Food and Animal Sciences Department in Texas A&M University campus. Samples are placed in an ice box for transportation to the lab. Using dissection tools, the aorta is separated from the heart and is cut into thoracic and abdominal segments. The thoracic aorta is further divided into upper descending and lower descending regions. These samples are stored at −20 °C in centrifuge tubes containing Phosphate buffer saline (PBS) solution.

Specimen Preparation

On the day of experiments, frozen specimens which are stored at −20 °C are thawed over a period of 12 h at 0 °C. Loose connective tissue is carefully removed without damaging the aortic layers. The aorta is cut and layed flat. Specimens of 12.5 mm diameter are punched out from the tissue. Around four samples per segment in both descending and ascending aorta are obtained.The specimens are stored in PBS at 0 °C until the time of testing. All specimens are tested within 4 h of thawing.

As much as possible, connective tissue is removed by teasing and cutting. The sample to be tested is taken out and dried with soft tissues.

Experiment Protocol

Next the specimen is attached to the rheometer plates. Cyanoacrylate adhesive is applied on the upper stainless steel attachment surface. The specimen's adventitial side is held pressed against upper surface for 30 s to adhere. Then, adhesive is applied on the lower surface and the sample is held at 1N for 5 min for complete adherence. This allows for a strong bond to form such that there won't be detachment during the test. When adhesion is improper the sample is discarded. The normal force is reduced to 0 N and PBS solution is added to create a bath at 25 °C. On addition of PBS, the normal force increases, which could be due to the sample getting hydrated. Before beginning the test, the normal force is brought back to 0.15N (Fig. 1).

The test is started with 50% shear strain amplitude cyclic at a constant shear rate. Ten cycles at 50, 40, 30, 20 and 10% shear strain amplitude are completed for each sample at two shear rates of 4% per second and 40% strain per second. The total test duration around 30 min for each sample. Around three to four samples are tested, both from the upper and lower thoracic aorta.

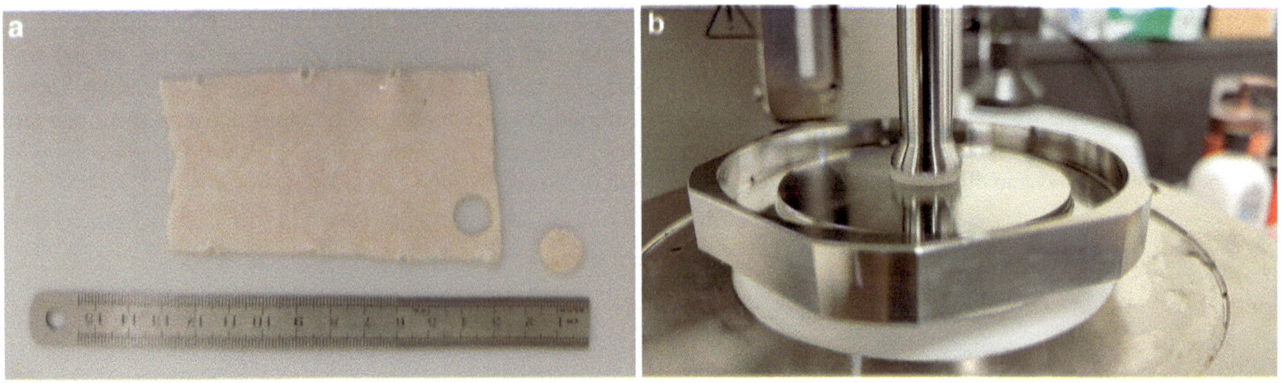

Fig. 1 (**a**) Aorta cut open and spread out with specimen punch holes. (**b**) Specimen on rheometer before introducing water bath

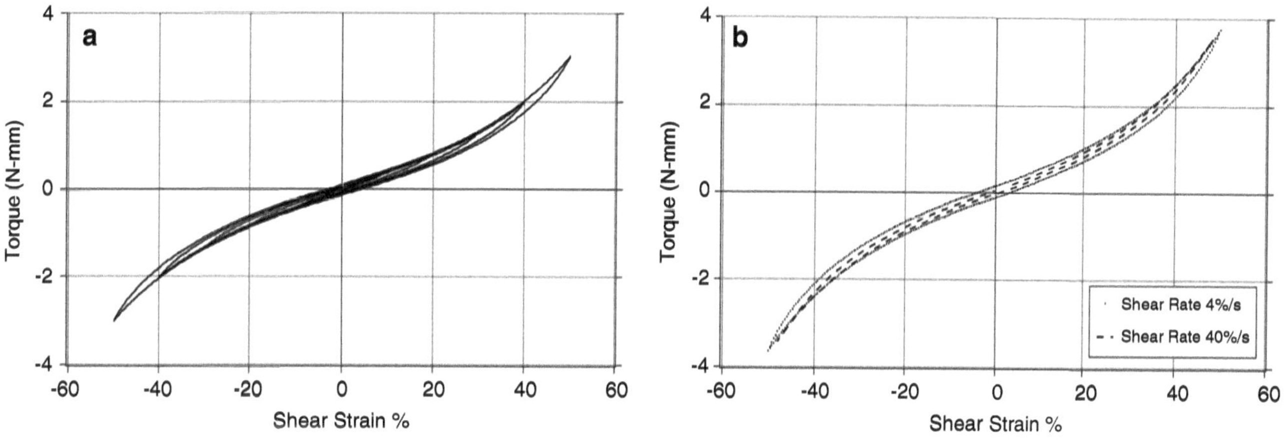

Fig. 2 (**a**) Torque vs Shear Strain for 10%, 20%, 30%, 40% and 50% Shear Sample 1. (**b**) Torque vs Shear Strain for two shear rates 4%/s and 40%/s

Results

Results from a shear test in which cyclic constant shear rate of 4%/s was applied is found in Fig. 2a. The final cycles from each amplitude of 10%, 20%, 30%, 40% and 50% applied shear strains can be seen here. In Fig. 2b we have torque-shear strain plots from the final cycle of the 50% shear strain amplitude response.

Discussion

From the response, it can be seen that the initial nearly linear response becomes nonlinear at around 20% shear strain. In one of the only other studies on shear on the aorta wall, Haslach et al. [5] translational shearing tests were conducted on bovine aortas and a linear relationship was found between shear strain and shear stress.

The results from our tests demonstrate a remarkable level of repeatability in the responses across specimens that is rarely seen in tests on biological tissues. A few characteristic traits of the moment-shear strain curves were common across the samples. The response was found to be almost identical under both positive and negative shear strain. The torque range was between 2 and 5 N-mm. Additionally, the torque-shear strain loops are overlapping for the different shear amplitudes. This was not observed by Dokos et al. [3] while testing porcine myocardium.

The normal force-shear strain response curves appear to be skewed in contrast to the torque-shear strain. This was observed across nearly all specimen with very few samples giving similar normal force responses for both positive and negative shear strain displacements.

It is important to note some of the limitations of this study. The samples were assumed to be of uniform thickness and consequently that the surfaces of the sample are completely adhered to the plates of the rheometer. We assume that the properties do not change throughout the test due to loss of hydration or exposure to the environment. The adhesive was considered not to diffuse too deeply into the tissue, although some diffusion is required for adhesion.

References

1. R. Avazmohammadi, D.S. Li, T. Leahy, E. Shih, J.S. Soares, J.H. Gorman, R.C. Gorman, M.S. Sacks, An integrated inverse model-experimental approach to determine soft tissue three-dimensional constitutive parameters: application to post-infarcted myocardium. Biomech. Model. Mechanobiol. **17**(1), 31–53 (2018). https://doi.org/10.1007/s10237-017-0943-1. http://link.springer.com/10.1007/s10237-017-0943-1
2. S.X. Deng, J. Tomioka, J.C. Debes, Y.C. Fung, New experiments on shear modulus of elasticity of arteries. Am. J. Physiol. Heart Circ. Physiol. **266**(35), H1–H10 (1994). https://www.physiology.org/doi/pdf/10.1152/ajpheart.1994.266.1.H1
3. S. Dokos, I.J. LeGrice, B.H. Smaill, J. Kar, A.A. Young, A triaxial-measurement shear-test device for soft biological tissues. J. Biomech. Eng. **122**(5), 471 (2000). https://doi.org/10.1115/1.1289624. http://biomechanical.asmedigitalcollection.asme.org/article.aspx?articleid=1403946

4. S. Dokos, B.H. Smaill, A.A. Young, I.J. LeGrice, Shear properties of passive ventricular myocardium. Am. J. Physiol. Heart Circ. Physiol. **283**(6), H2650–H2659 (2002)
5. H.W. Haslach, L.N. Leahy, P. Fathi, J.M. Barrett, A.E. Heyes, T.A. Dumsha, E.L. McMahon, Crack propagation and its shear mechanisms in the bovine descending aorta. Cardiovasc. Eng. Technol. **6**(4), 501–518 (2015). https://doi.org/10.1007/s13239-015-0245-7. http://link.springer.com/10.1007/s13239-015-0245-7
6. V.A. Kasyanov B.A. Purinya, E.T. Determination of the shear modulus of human blood vessel walls. Polym. Mech. (2) (1979). https://doi.org/10.1152/ajpheart.1994.266.1.H1. https://link.springer.com/content/pdf/10.1007/BF00860800
7. X. Lu, J. Yang, J.B. Zhao, H. Gregersen, G.S. Kassab, Shear modulus of porcine coronary artery: contributions of media and adventitia. Am. J. Physiol. Heart Circ. Physiol. **285**(5), H1966–H1975 (2003). https://doi.org/10.1152/ajpheart.00357.2003. http://www.physiology.org/doi/10.1152/ajpheart.00357.2003
8. G. Sommer, M. Eder, L. Kovacs, H. Pathak, L. Bonitz, C. Mueller, P. Regitnig, G.A. Holzapfel, Multiaxial mechanical properties and constitutive modeling of human adipose tissue: a basis for preoperative simulations in plastic and reconstructive surgery. Acta Biomater. **9**(11), 9036–9048 (2013). https://doi.org/10.1016/J.ACTBIO.2013.06.011. https://www.sciencedirect.com/science/article/pii/S174270611300295X
9. G. Sommer, A.J. Schriefl, M. Andrä, M. Sacherer, C. Viertler, H. Wolinski, G.A. Holzapfel, Biomechanical properties and microstructure of human ventricular myocardium. Acta Biomater. **24**, 172–192 (2015). https://doi.org/10.1016/j.actbio.2015.06.031. http://dx.doi.org/10.1016/j.actbio.2015.06.031
10. A.T.J. Vossoughi, Determination of an effective shear modulus of aorta. Russ. J. Biomech. **1**, 20–35 (1998)

Imaging of the Scattering of High-Intensity Focused Ultrasonic Waves at Artificial Bone Replicas

Matthew Brown, David Sanford, and Christoph Schaal

Abstract High-intensity focused ultrasound is a technology currently used to treat bodily tissue for various medical purposes. For example, it has been applied for tissue ablation as a treatment for prostate cancer. However, the effective targeting of tissue deeper inside the body remains challenging because bones obstruct and scatter ultrasonic waves, which reduces the energy transmission to the desired location. Thus, understanding wave-scattering effects on focal point location and intensity is crucial for the expansion of the technology. Previous ultrasound visualization studies have not examined these effects in detail, especially for curved bone geometries.

Hence, in this work, the effects of bones on the transmission of focused ultrasound is investigated for hollow and solid cylindrical bone geometries. In laboratory experiments, images of wave fields are captured using shadowgraph techniques. The method uses a pulsed laser synced with a CMOS camera. Ultrasonic waves cause periodic local changes in density of the water, producing bright and dark patterns of laser transmission corresponding to the wave peaks and troughs. Improvements in the experimental setup and image processing compared to previous work allow for an expansion of the field of view with higher contrast. The bone replicas scatter the ultrasonic wave field and the effect of obstruction on ultrasound focal point intensities is quantified using pixel intensity measurements. In addition to visualization, differences in the pressure fields are recorded using a hydrophone and compared to the results obtained from the shadowgraph images. This experimental work provides a reference for future research in medical ultrasound and has the potential to lead to the development of methods that optimize the targeting of tissue deep in the human body.

Keywords High-intensity focused ultrasound · Wave scattering · Bones · Shadowgraphy · Hydrophone

Introduction

This work is motivated by the emergence of focused ultrasound as a method of therapeutic treatment for a variety of disorders and ailments in the body. Recently, the technology has seen increased attention and is at various stages of clinical development for use in the ablation of cancer cells [1–3], targeted drug delivery [4, 5], and for the removal of blood clots that cause strokes [6]. However, these methods are generally limited to applications in homogeneous tissues near the surface of the body since bone obstructions scatter incident waves [7–10].

In its current state, high-intensity focused ultrasound (HIFU) technology incorporates focusing and guidance of ultrasonic waves [11] as well as real-time power control [12]. Both methods improve the accuracy of HIFU applied to homogeneous tissue. However, clinical applications built on computer modeling of multi-layer media [7, 9, 13] or with bone obstructions [10] are elusive. Numerical studies have approached the modeling of propagation in two dimensions [8], in three dimensions with scattering [10], and in the study of tissue energy absorption [9]. Experiments for studying the interactions of HIFU with obstructions [14] show variable (unpredictable) energy deposition in the target tissue when ultrasonic waves have penetrated a flat bone. Similarly, Saito et al. [15] previously showed transmittance fluctuation through a flat plate. The obstruction effects for simple bone geometries have also been investigated by Brown et al. [13] and Schaal and Durgesh [16], whose findings

M. Brown (✉) · D. Sanford
Department of Mechanical Engineering, California State University, Northridge, CA, USA
e-mail: matthew.brown.279@my.csun.edu; david.sanford.707@my.csun.edu

C. Schaal
Department of Mechanical Engineering, California State University, Northridge, CA, USA

Mechanical and Aerospace Engineering Department, University of California, Los Angeles, CA, USA
e-mail: christoph.schaal@csun.edu

M. E. Grady (ed.), *Mechanics of Biological Systems and Materials & Micro-and Nanomechanics, Volume 4*,
Conference Proceedings of the Society for Experimental Mechanics Series, https://doi.org/10.1007/978-3-030-30013-5_8

suggest that wave-scattering at a flat plate produces areas of increased intensity in nearby tissue. However, for the application of the technique to complex, curved bone regions such as the spinal column for nerve pain therapy [17], scattering effects have scarcely been investigated. To evaluate numerical or other test results, shadowgraph wave-visualization techniques can be used. The shadowgraph method (similar to schlieren visualization) has been historically used in the study of fluid dynamics, for example to create images of shock-wave profiles around a supersonic aircraft. More applications are presented in the work of Settles [18] and Settles and Hargather [19]. More recently, schlieren and shadowgraph techniques have been used to the visualize the density gradients produced by HIFU in water [13, 20–22], thus facilitating the study of wave-scattering by bones.

In this work, shadowgraph images are presented for flat and curved bone geometries located in the wave path produced by a HIFU transducer. Pixel values are used to extract intensities of the pressure fields. Hydrophone measurements are used to verify and compare with shadowgraph image results without obstructions, and verify previous work on the relocation of peak intensities for partial obstruction by a flat plate.

Methods

Shadowgraph Method and Image Processing

The shadowgraph visualization technique is used in this study because it is simple and cost-efficient compared to the well-known schlieren method. A shadowgraph setup can still be used to achieve high-quality images of ultrasound wave fields [13, 23]. Both methods rely on the principle that the presence of a density gradient in a transparent fluid refracts incident light. In this case, HIFU waves produce a pressure gradient sufficient to bend incident light into a bright and dark pattern. However, because shadowgraph setups do not use a knife edge to cut off and remove background light, image post-processing is needed to achieve high-contrast images. First, multiple images for a given time step are averaged to prevent air bubbles or dust particles from obscuring pixel information. Then, digital image background subtraction is performed to remove undeflected light and enhance contrast, similar to the function of the knife edge in the schlieren technique. Finally, low-pass filtering is applied to reduce noise in the images. This process is depicted in Fig. 1, where the raw image is shown in Fig. 1a, the background image in Fig. 1b and the post-processed image in Fig. 1c. The field of view (FOV) is limited to approx. 25.5 mm due to the experimental setup as described below.

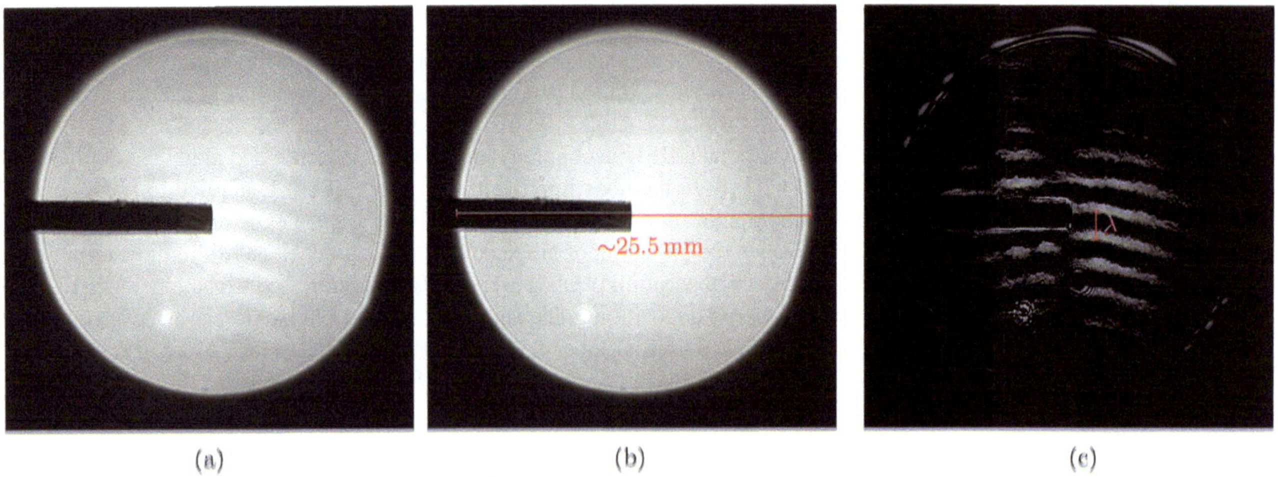

Fig. 1 Post-processing of shadowgraph images: averaging, image subtraction, and low-pass filtering are performed to enhance contrast and improve the signal-to-noise ratio. (**a**) Raw image. (**b**) Background image. (**c**) Post-processed image

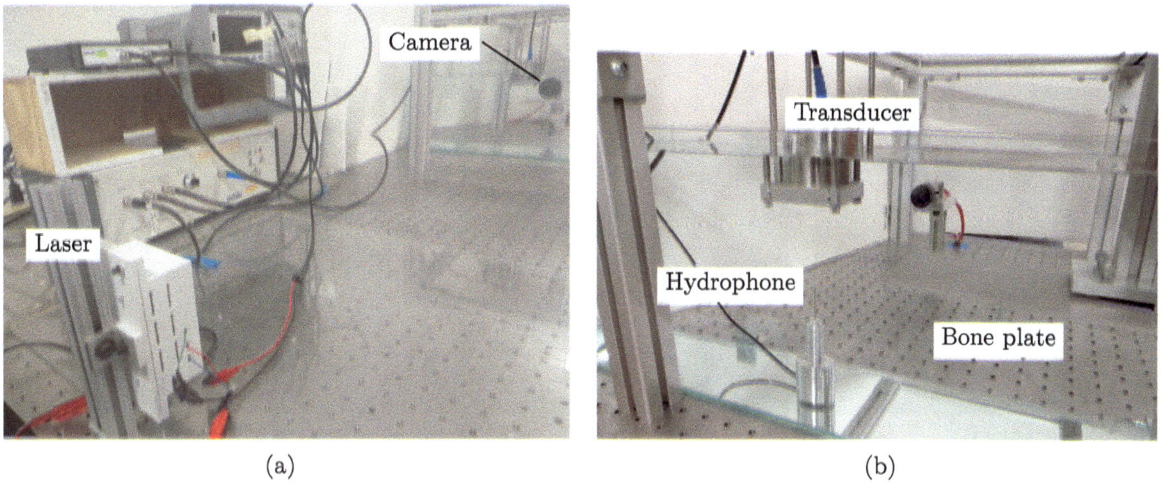

Fig. 2 Laboratory setup: imaging system to visualize focused ultrasonic waves and their scattering at bone tissue as well as setup used for hydrophone measurements. (**a**) Visualization system. (**b**) Hydrophone measurement setup

Experimental Setup and Hydrophone Measurements

For both the acquisition of shadowgraph images and hydrophone measurements, an experimental setup on a vibration-isolated optical bench is used as shown in Fig. 2. A multi-stage positioning system controls the position of the HIFU transducer (Precision Acoustics PA510). The position of the 2 mm thick bone plate specimen (Sawbones) that is used in some experiments is also controlled with the linear positioning system. The 10 mm outer diameter cylindrical bones (hollow with 2 mm wall thickness and solid) are suspended in the water using a 0.05 mm diameter filament.

For the shadowgraph experiments, a pulsed laser diode (Osram SPL LLP90) with a wavelength of 905 nm is used. Because of the fast-traveling ultrasonic waves (approx. 1500 m/s), a strobe method is implemented through the use of a custom-made control software in Matlab. With the test environment blacked out, the 2048 × 1088 pixel CMOS camera (Basler Aca2000-50 gm) is triggered (opening the shutter) at the same time as the signal generator (Keysight 33512B). The signal is amplified by an RF amplifier (Ritec 3000A) that is driving the HIFU transducer at 670 kHz center frequency. After a short delay based on the time that the ultrasound must travel before entering the illuminated FOV, the 30 ns laser pulse is triggered. Lastly, the camera shutter is closed, and the image is saved. This process is repeated at subsequent time increments to acquire images that show the entire propagation of the ultrasonic waves through the FOV, which is limited to about 25.5 mm for the camera lens used (Nikkor AF-S 50 mm f/1.4).

The hydrophone experiments are run independent of the shadowgraphy to avoid image obstruction by the hydrophone (Müller-Platte needle probe). In Fig. 2b, the machined hydrophone base is seen on the floor of the tank holding the needle hydrophone vertically. To accurately position the hydrophone tip at the HIFU focal point, the base is held fixed while the HIFU transducer is visually centered. This position is then refined by finding the largest amplitude from an unobstructed hydrophone signal. This baseline position is used as a reference for all experiments. The linear stages are used to reposition the system within $y \in [30, 80]$ mm in the vertical direction, where $y = 0$ mm is the transducer surface and $y = 73.19$ mm is the theoretical focal point. In the horizontal direction, experiments are limited to $x \in [-8, 8]$ mm, where $x = 0$ mm is aligned with the center of the transducer. Through additional experiments, it is found that insignificant signal strength occurs beyond the x-limits. At the focal points, induced waves packets arrive at about 49 µs and pass this point at about 80 µs. Hence, the intensities of the waves are determined through sum of squares of the time histories within these boundaries.

Results

Unobstructed HIFU Wave Propagation

As a reference for obstructed cases, the ultrasound pressure field generated by the HIFU transducer is captured without an obstruction. Images with the center of the FOV aligned with the focal point of the transducer are shown in Fig. 3. As time increases, the focusing effect of the transducer is apparent, and the pixel intensity increases as the waves approach the focal

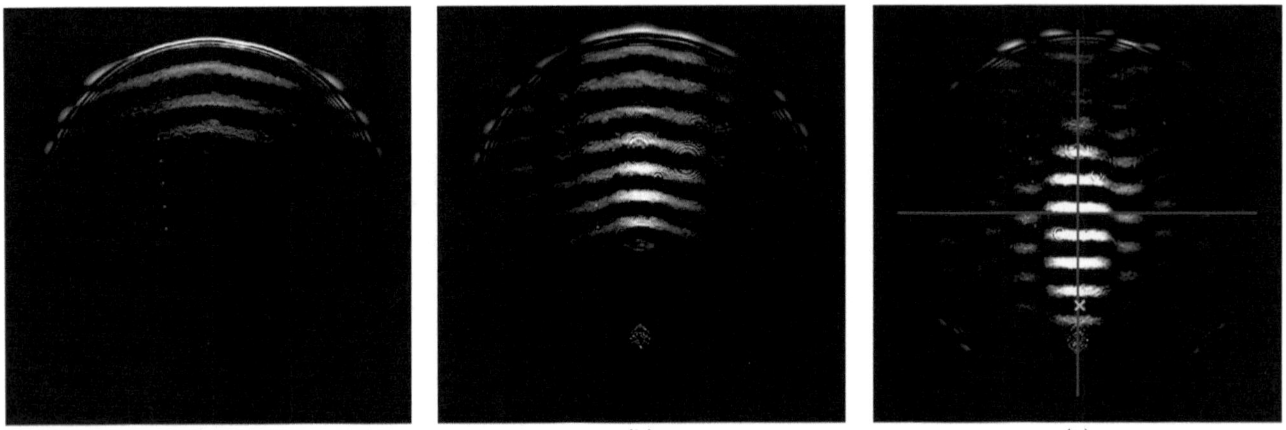

Fig. 3 Shadowgraph images captured at the focal point of the HIFU transducer at various time steps (unobstructed). (**a**) $t = 45.31\,\mu s$. (**b**) $t = 50.31\,\mu s$. (**c**) $t = 55.31\,\mu s$

point. For display, each image is normalized to the maximum pixel intensity in this unobstructed case, so that qualitative comparisons between all presented images can be made. Even with image averaging, background subtraction and filtering, it can be seen that pixel distortion occurs in shadowgraph images due to contamination by air bubbles and dust particles. Additionally, a bright and dark fringe pattern can be seen at the edges of the FOV. For quantitative pixel intensity evaluation, these regions are excluded. Finally, laser glare appears in the bottom of some images. Despite these anomalies, meaningful information can be extracted for the unobstructed and obstructed cases presented in the remainder of this work. For example, the wavelength of the HIFU transducer is experimentally derived from unobstructed shadowgraph images, such as Fig. 1, to be $\lambda \approx 1.95$ mm. This results in an approx. 11 % error for a sound wave speed $c = 1484.8$ m/s in water at 20.7 °C [24].

To validate the use of shadowgraph pixel intensity in quantifying wave field intensities, hydrophone measurements are performed in the focal region of the HIFU transducer and compared to shadowgraph intensities. Measurements along the vertical center line of the transducer through the focal point, shown as a blue vertical line in Fig. 3c, are maximum-normalized and presented in Fig. 4a. Pixel data is extracted by taking the maximum intensity across all captured images for selected pixels. The apparent noise in pixel intensity (shown as solid lines) is smoothed using a moving average filter to allow for a better comparison with the hydrophone data (+). While the curves do not match, the trends are similar and the peak locations at $y = 73.71$ mm and $y = 72.63$ mm for the shadowgraph and hydrophone measurement, respectively, are near the theoretical focal point of $y = 73.19$ mm that is provided by the manufacturer. Next, the induced HIFU waves are measured along a horizontal line intersecting the focal point, shown as a red horizontal line in Fig. 3c. The maximum-normalized results are presented in Fig. 4b. The shadowgraph (solid line) and hydrophone measurements ($\times$) are in good agreement and show a concentration of energy to a region of $x \in [-2, 2]$ mm.

HIFU Wave-Scattering at Bone Plate Replica

A 2 mm flat bone replica plate is introduced at $D = 20$ mm above the focal point, blocking the ultrasound path with $o = 50\,\%$ level of obstruction, in order to capture wave-scattering for partial obstruction. A sequence of shadowgraph images can be seen in Fig. 5, where the images are centered to the plate and the vertical center-line of the HIFU transducer. The focal point is not captured in these images due to the limited FOV. The partial blockage of the ultrasound path produces significant changes in the shape of the HIFU waves. From a visual, qualitative analysis, it can be seen that the waves have reduced pixel intensity, compared to the bright white waves captured at the focal point without obstruction (see Fig. 3). This low intensity region is desirable in HIFU applications, so that tissue along the path to the target area remains (nearly) unaffected by the HIFU treatment. It can be seen that the partial obstruction causes splitting and a phase shift of the transmitted ultrasound beams in Fig. 5c. This not only reduces the focusing effect, but it also produces two regions of higher intensities away from the central axis of the transducer.

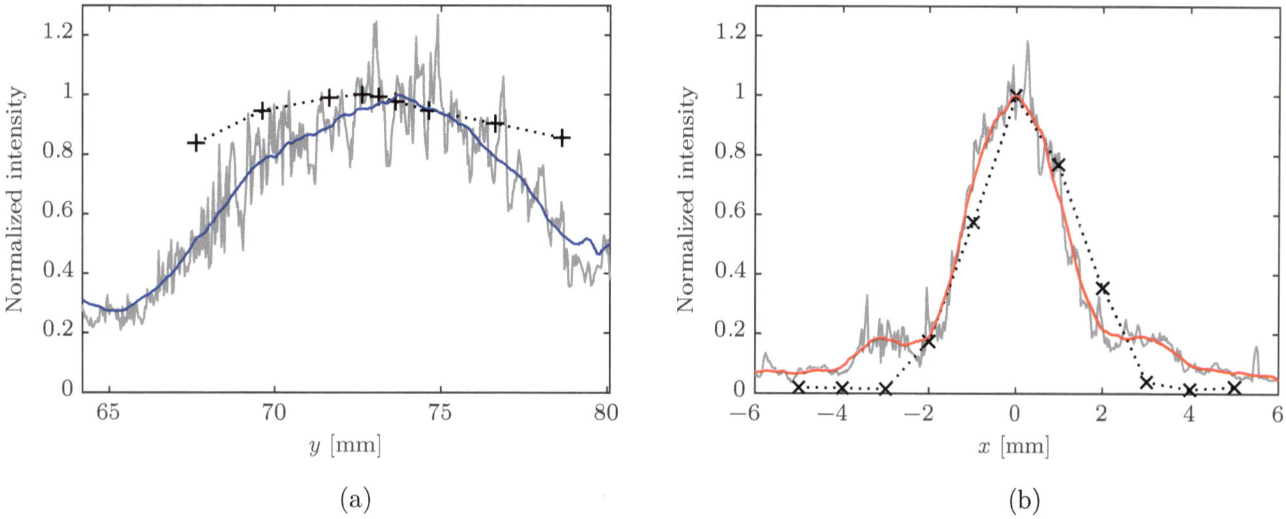

Fig. 4 Comparison of shadowgraph intensities with hydrophone measurements near the focal point. Shadowgraph intensities are smoothed using a moving average filter. (**a**) Shadowgraph data is shown as a solid line, hydrophone measurements as $+$. (**b**) Shadowgraph data is shown as a solid line, hydrophone measurements as $\times$

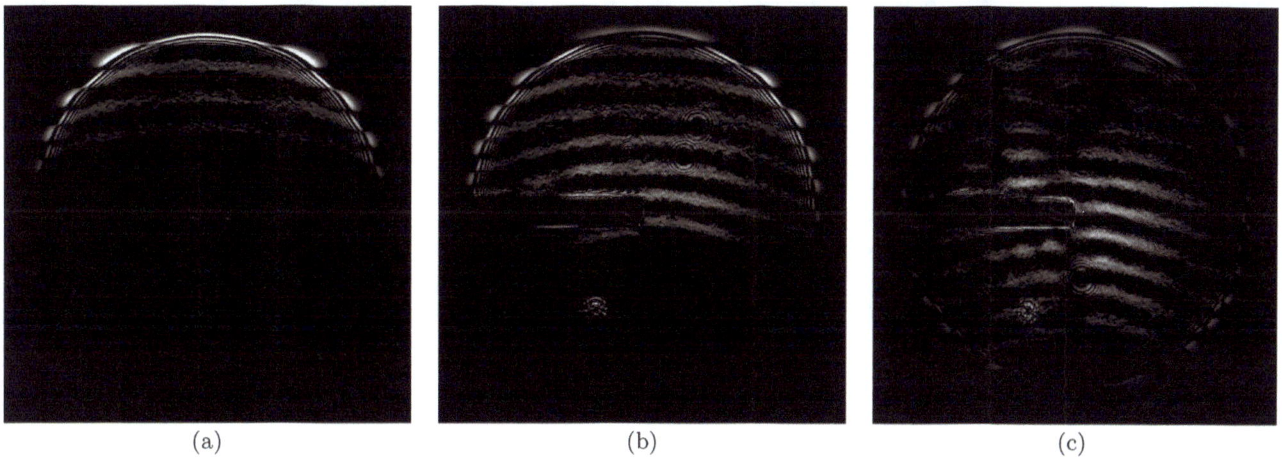

Fig. 5 Shadowgraph images captured near bone replica at various time steps (plate at $o = 50\,\%$ and $D = 20\,\text{mm}$). (**a**) $t = 31.76\,\mu\text{s}$. (**b**) $t = 36.76\,\mu\text{s}$. (**c**) $t = 41.76\,\mu\text{s}$

These observations can be confirmed with hydrophone measurements. The measurements are recorded along a horizontal line crossing the focal point of the HIFU transducer, where the focal point is at $x = 0\,\text{mm}$. The intensities captured for the unobstructed (red $\times$) and plate-obstructed (blue $\circ$) cases are plotted in Fig. 6. The maximum recorded intensities from the obstructed case is reduced to about 23 %, and the main lobe is transformed into two side lobes with peaks at about 2 mm from the center. Hence, this confirm that the obstructed wave intensity is shifted laterally away from the center and diminished relative to the unobstructed case.

HIFU Wave-Scattering at Curved Bone Replicas

As shown for the bone plate case, even the obstruction by relatively simple geometries may cause a relocation of maximum intensities in the wave field. To inform applications in areas with curved bone geometry obstruction such as for the spinal column, shadowgraph images are taken for two curved bone replicas located at the focal point ($D = 0\,\text{mm}$): a hollow cylinder specimen (10 mm outer diameter and 2 mm wall thickness) and a solid cylinder (10 mm outer diameter). A sequence of shadowgraph images for the hollow cylinder can be seen in Fig. 7. For these images, as well as those for the solid cylinder,

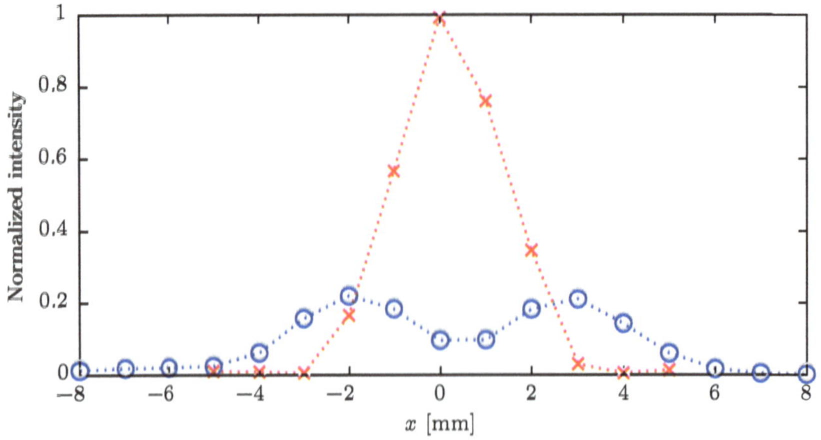

Fig. 6 Maximum amplitudes captured for unobstructed (*cross*) and plate obstructed (*opencircle*) hydrophone measurements. A bone replica plate is introduced at $D = 20$ mm and $o = 50\,\%$ for the obstructed case

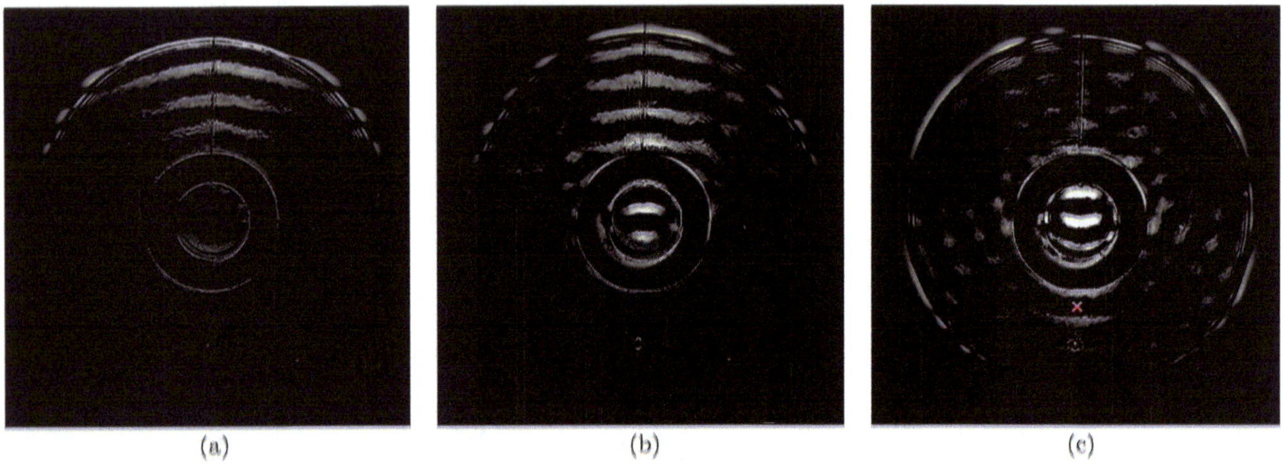

Fig. 7 Shadowgraph images captured near bone replica at various time steps (hollow cylinder at focal point). (**a**) $t = 45.31\,\mu$s. (**b**) $t = 50.31\,\mu$s. (**c**) $t = 55.31\,\mu$s

the thin filament used to suspend the specimens can be seen as a thin dark vertical line, which removes vertical center-line information from the images. As in the unobstructed case, the pixel intensity increases significantly as the waves approach the focal point, which is coincident with the center of the cylinder. Ring patterns appear as outward rippling of scattered waves and are best visible in Fig. 7c. Transmitted waves with reduced intensity can be seen as gray fringes below the hollow cylinder.

Shadowgraph images are also captured for a solid cylinder at the HIFU focal point. The images are recorded at the same time steps as those for the hollow cylinder a sequence of three images is shown in Fig. 8. As mentioned above, pixel intensities can directly be compared across the different cases. It can be seen that the waves transmitted through the solid specimen have slightly lower pixel values than those transmitted through the hollow cylinder. This can be attributed to the fact that more energy is scattered and absorbed by the solid bone replica. Finally, it appears that waves traveling through the solid cylinder undergo a more significant increase in velocity than in the hollow case because of the longer propagation path through bone-mimicking material and the higher density of the specimen relative to water.

These observations about the transmitted waves are verified quantitatively by extracting pixel intensities just below the cylinders, marked as × in the corresponding shadowgraph images. The intensities from the hollow (red dashed line) and solid cylinder (green dash-dotted line) cases are normalized to the unobstructed case (solid gray line), and the results are shown in Fig. 9. It should be noted that the curves are not perfectly smooth as the images are taken in steps of 0.5 μs, which results in a sampling frequency of 2 MHz. Nonetheless, it can be clearly seen that the wave pulse transmitted through the solid cylinder arrives before the one crossing the hollow cylinder bone replica, while the unobstructed wave arrives last. Additionally, the

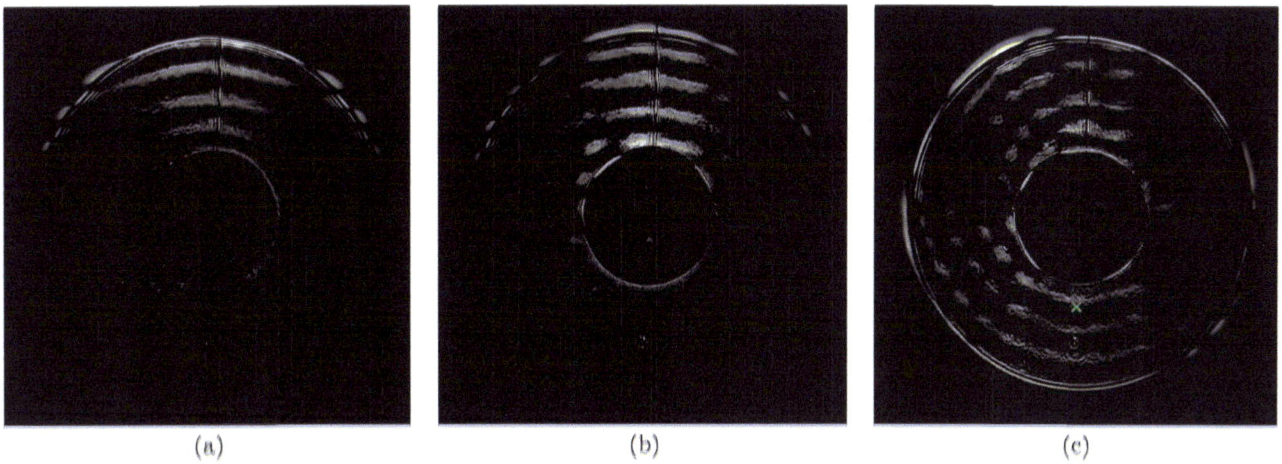

Fig. 8 Shadowgraph images captured near bone replica at various time steps (solid cylinder at focal point). (**a**) $t = 45.31\,\mu s$. (**b**) $t = 50.31\,\mu s$. (**c**) $t = 55.31\,\mu s$

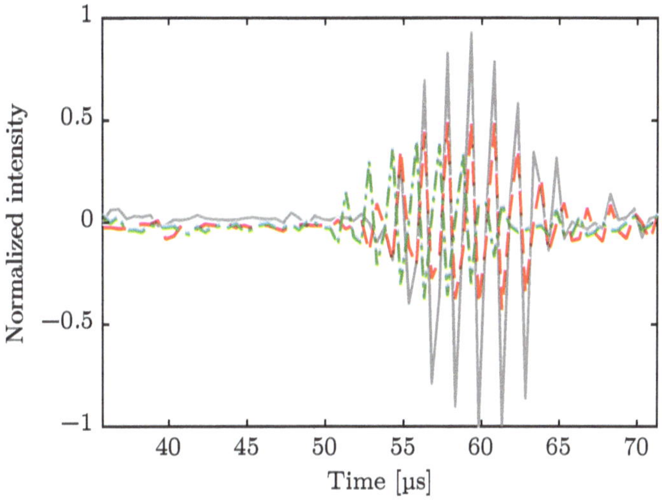

Fig. 9 Shadowgraph pixel intensity time histories measured at a point below the solid cylinder (green dash-dotted line) and hollow cylinder (red dashed line). The unobstructed case is shown as solid gray line

waves transmitted through the bone specimens both show significantly reduced amplitudes, with the solid cylinder causing the greatest reduction.

Conclusions

Shadowgraph imaging is an effective ultrasound wave visualization technique. The presented images produce qualitative data about the location and relative intensity of induced high-intensity focused ultrasonic waves. Furthermore, several obstruction cases are also imaged, showing in great detail the scattered wave patterns: for a bone plate replica at $D = 20\,\text{mm}$ and $o = 50\,\%$, as well as solid and hollow bone cylinder replicas at the focal point. Hydrophone measurements confirm the general distribution and intensities of waves in the unobstructed and plate-obstruct cases. Image processing allows for the extraction of quantitative data in form of pixel intensities. It is shown that both results from both methods are generally in good agreement, thus increasing the confidence in directly using the shadowgraph images for quantitative analyses.

Introducing smaller, curved geometries at the HIFU focal point present challenges in form of suspending the obstructing object in such a way that the fixture has a minimal effect on the captured shadowgraph images. Nonetheless, it is shown that focused waves partially propagate through the bone material and reach the original focal point at the center of the cylindrical bone specimens as well as the area below. It is also shown that the shadowgraph pixel intensity analysis can be very effective to quantitatively analyze an entire wave field, possibly in two dimensions in future works. This and additional future analyses might lay the groundwork for new clinical HIFU application in the presence of bone obstructions.

Acknowledgements The RF amplifier provided by Dr. Ajit Mal (UCLA) is gratefully acknowledged.

References

1. G. Ter Haar, Therapeutic ultrasound. Eur. J. Ultrasound 9(1), 3–9 (1999). https://doi.org/10.1016/S0929-8266(99)00013-0
2. J.E. Kennedy, High-intensity focused ultrasound in the treatment of solid tumours. Nat. Rev. Cancer **5**(4), 321–327 (2005). https://doi.org/10.1038/nrc1591
3. Y.-F. Zhou, High intensity focused ultrasound in clinical tumor ablation. World J. Clin. Oncol. **2**(1), 8 (2011). https://doi.org/10.5306/wjco.v2.i1.8
4. K. Ferrara, R. Pollard, M. Borden, Ultrasound microbubble contrast agents: fundamentals and application to gene and drug delivery. Annu. Rev. Biomed. Eng. **9**, 415–447 (2007). https://doi.org/10.1146/annurev.bioeng.8.061505.095852
5. S. Dromi, V. Frenkel, A. Luk, B. Traughber, M. Angstadt, M. Bur, J. Poff, J. Xie, S. K. Libutti, K. C. Li, et al., Pulsed-high intensity focused ultrasound and low temperature–sensitive liposomes for enhanced targeted drug delivery and antitumor effect. Clin. Cancer Res. **13**(9), 2722–2727 (2007). https://doi.org/10.1158/1078-0432.CCR-06-2443
6. Z. Wang, T. Komatsu, H. Mitsumura, N. Nakata, T. Ogawa, Y. Iguchi, M. Yokoyama, An uncovered risk factor of sonothrombolysis: substantial fluctuation of ultrasound transmittance through the human skull. Ultrasonics **77**, 168–175 (2017). https://doi.org/10.1016/j.ultras.2017.02.012
7. F. Feng, A. Mal, M. Kabo, J. C. Wang, Y. Bar-Cohen, The mechanical and thermal effects of focused ultrasound in a model biological material. J. Acoust. Soc. Am. **117**(4), 2347–2355 (2005). https://doi.org/10.1121/1.1873372
8. P.V. Yuldashev, W. Kreider, O.A. Sapozhnikov, N. Farr, A. Partanen, M.R. Bailey, V. Khokhlova, Characterization of nonlinear ultrasound fields of 2d therapeutic arrays, in *Proceedings of IEEE International Ultrasonics Symposium*, vol. 2012 (NIH Public Access, 2012), p. 1. https://doi.org/10.1109/ULTSYM.2012.0231
9. D. Christensen, S. Almquist, Incorporating tissue absorption and scattering in rapid ultrasound beam modeling, in *Energy-based Treatment of Tissue and Assessment VII*, vol. 8584 (International Society for Optics and Photonics, 2013), p. 85840X. https://doi.org/0.1117/12.2008021
10. P. Gélat, G. Ter Haar, N. Saffari, HIFU scattering by the ribs: constrained optimisation with a complex surface impedance boundary condition, in *Journal of Physics: Conference Series*, vol. 498 (IOP Publishing, Bristol, 2014), p. 012004. https://doi.org/10.1088/1742-6596/498/1/012004
11. J.K. Enholm, M.O. Kohler, B. Quesson, C. Mougenot, C.T. Moonen, S.D. Sokka, Improved volumetric MR-HIFU ablation by robust binary feedback control. IEEE Trans. Biomed. Eng. **57**(1), 103–113 (2010). https://doi.org/10.1109/TBME.2009.2034636
12. T. Long, V. Amin, S. McClure, R. Robert, L. Wu, R. Thompson, T. Ryken, P3C-8 Techniques for Real-Time Monitoring and Control for HIFU (High Intensity Focused Ultrasound) Ablationin Porcine Brains In Vitro Studies, in *Proceeding of IEEE Ultrasonics Symposium, 2007* (IEEE, Piscataway, 2007), pp. 1788–1791. https://doi.org/10.1109/ULTSYM.2007.450
13. M. Brown, M. Safisamghabadi, C. Schaal, V. Durgesh, Visualization of the scattering of focused ultrasonic waves at solid-fluid interfaces. Proc. SPIE **10600**, 10600–10610 (2018). https://doi.org/10.1117/12.2296744
14. R. Omura, Y. Shimazaki, S. Yoshizawa, S.-i. Umemura, Quantitative measurement of focused ultrasound pressure field using subtraction shadowgraph. Jpn. J. Appl. Phys. **50**(7S), 07HC07 (2011). https://doi.org/10.1143/jjap.50.07hc07
15. O. Saito, Z. Wang, H. Mitsumura, T. Ogawa, Y. Iguchi, M. Yokoyama, Substantial fluctuation of acoustic intensity transmittance through a bone-phantom plate and its equalization by modulation of ultrasound frequency. Ultrasonics **59**, 94–101 (2015). https://doi.org/10.1016/j.ultras.2015.01.017
16. C. Schaal, V. Durgesh, Investigation of the scattering of focused ultrasonic waves at bones, in *Proceedings of the ASME International Mechanical Engineering Congress and Exposition—IMECE 2018*, Pittsburgh (2018). https://doi.org/10.1115/IMECE2018-87133
17. S. Harnof, Z. Zibly, L. Shay, O. Dogadkin, A. Hanannel, Y. Inbar, I. Goor-Aryeh, I. Caspi, Magnetic resonance-guided focused ultrasound treatment of facet joint pain: summary of preclinical phase. J. Ther. Ultrasound **2**(1), 9 (2014). ISSN 2050-5736. https://doi.org/10.1186/2050-5736-2-9
18. G. Settles, *Schlieren and Shadowgraph Techniques: Visualizing Phenomena in Transparent Media* (Springer, Berlin, 2001). ISBN 978-3-540-66155-9. https://doi.org/10.1007/978-3-642-56640-0
19. G.S. Settles, M.J. Hargather, A review of recent developments in schlieren and shadowgraph techniques. Meas. Sci. Technol. **28**(4), 042001 (2017). https://doi.org/10.1088/1361-6501/aa5748
20. N. Kudo, H. Ouchi, K. Yamamoto, H. Sekimizu, A simple schlieren system for visualizing a sound field of pulsed ultrasound. J. Phys. Conf. Ser. **1**, 146–149 (2004). https://doi.org/10.1088/1742-6596/1/1/033
21. N. Kudo, A simple technique for visualizing ultrasound fields without schlieren optics. Ultrasound Med. Biol. **41**(7), 2071–2081 (2015). https://doi.org/10.1016/j.ultrasmedbio.2015.03.004
22. T. Neumann, H. Ermert, Schlieren visualization of utlrasonic waves with high spatial resolution. Ultrasonics **44**, 1561–1566 (2006). https://doi.org/10.1016/j.ultras.2006.05.209
23. N. Kudo, A simple technique for visualizing ultrasound fields without schlieren optics. Ultrasound Med. Biol. **41**(7), 2071–2081 (2015). https://doi.org/10.1016/j.ultrasmedbio.2015.03.004
24. M. Greenspan, C.E. Tschiegg, Tables of the speed of sound in water. J. Acoust. Soc. Am. **31**(1), 75–76 (1959). https://doi.org/10.1121/1.1907614

A Comparison Between Bearing and Non-bearing Human Bone: Mechanical Testing and Micro-Architecture Assessment

Xavier Roothaer, Rémi Delille, Hervé Morvan, Eric Markiewicz, and Christian Fontaine

Abstract An adult skeleton has over 200 bones. Each bone has a specific function. These different functions bring different types of mechanical loadings. To ensure a healthy behavior of the bone, a throughout life process occurs on the micro-architecture of the cortical bone. This process highly depends on the stress applied on the bone. The micro-architecture is able to supply blood and nutrients into the bone matrix and acts on the bone remodeling process. The architecture is formed by vascular canals where the orientation depends on the main axis of the mechanical loading. Vascular canals of the cortical bone located in long bone diaphysis are mainly oriented along the longitudinal axis of the diaphysis. Moreover, the canal network is tortuous and several geometrical features should affect the macroscopic behavior of the bone. The novelty of this work is to use a method which accurately describes the vascular architecture of the cortical bone based on micro-computed tomography and to correlate these results with the macroscopic mechanical behavior of the cortical bone. This experimental study is based on the extraction of cortical bone samples from four cadaveric human subjects. For each human subject, left and right humeri and femurs are studied. Dumbbell-shaped bone specimens are prepared from each bone. Care was taken to preserve bone properties after extraction: a maximum of 15 days was fixed between the extraction and the mechanical testing. Samples were constantly kept hydrated and stored at 4 °C in order to avoid frost/defrost cycles. Specimens are scanned and subsequently mechanically loaded to failure. Results will show the impact of the laterality of the bone on the architecture, the impact of loading on the bone on the architecture by comparing humeral and femoral samples for each human subject and finally the impact of the architecture on the mechanical behavior of the bone.

Keywords Cortical bone · Architecture · Tensile test · Bearing bone · Haversian canals

Introduction

Long bones are located in the human appendicular skeleton. These bones have a predominant central shaft which is called diaphysis. On each ending, the epiphysis marks out the shaft. This is the part which ensures the articulation with the adjacent bones. Consequently, it is coated by an articular cartilage. However, due to this function, the portion needs to be tremendously vascularized and its architecture is akin to a spongy material. Hence, in long human bone the mechanical behavior is warranted by the diaphysis which is mainly covered by a few millimeters shell named cortical bone (or compact bone). The cortical bone has a highly hierarchized micro-architecture so as to carry blood vessels and nutrients inside the bone. From a mechanical point of view, the macroscopic behavior can be approximated by similar methods as used for synthetic materials (Long or short fibers reinforced materials). Indeed, micro-computed tomography is often used for this type of material. The device helps to quantify the architecture without destroying material (NDT). Moreover, since 2000s, it became the gold-standard method for cortical bone [1]. Nevertheless, most of human body FEM models use a generic material law for the entire cortical bone of the skeleton. In the 60s, Frost showed a close link between mechanical stimuli and a dynamic process of bone remodeling [2, 3]. This biological process means that the cortical bone is able to adapt itself according the commonly applied stresses. Consequently, cortical bone architecture differs from one bone to another. Many studies assess cortical bone by harvesting from animals (mostly bovine) or human femur [4–7]. This choice is mainly explained by the relative ease of sample extraction (thick cortical shell). However, it brings a lack of data for other bones. For

X. Roothaer (✉) · R. Delille · H. Morvan · E. Markiewicz
CNRS, UMR 8201, LAMIH, University Polytechnique Hauts-de-France, Valenciennes, France
e-mail: xavierroothaer@gmail.com; remi.delille@uphf.fr; herve.morvan@uphf.fr; eric.markiewicz@uphf.fr

C. Fontaine
Department of Anatomy, Faculty of Medicine, University of Lille 2, Lille Cedex, France
e-mail: Christian.FONTAINE@CHRU-LILLE.FR

M. E. Grady (ed.), *Mechanics of Biological Systems and Materials & Micro-and Nanomechanics, Volume 4*,
Conference Proceedings of the Society for Experimental Mechanics Series, https://doi.org/10.1007/978-3-030-30013-5_9

instance, non-bearing bones are poorly investigated. Nevertheless, it is obvious that stresses applied on these bones highly differ from the femur. Some studies carried out an assessment of the humerus but are still limited [8–10]. Furthermore, the cortical bone is lifestyle, age and gender dependent [8, 11]. So as to obtain comparative data between two different bones, the harvest should be carried out on same human subject. The purpose of the study is to assess the mechanical behavior of both bearing and non-bearing bones by focusing on human humerus and femur, respectively. Dumbbell-shaped samples were extracted so as to be mechanical tested until failure. Tensile tests were carried out. Each sample was previously scanned using a μ-CT device. The originality of this work is the use of a Python script which analyzes tomographic slices in order to compute three-dimensional geometrical features of the vascular network located within the bone [12, 13] whereas other current studies are limited to in-plane analysis [4, 14]. Numerous geometrical features of canals are individually computed and make correlations possible with mechanical behavior.

Materials and Method

The experimental study is based on the harvest of humeri and femurs from 4 cadaveric male subjects with an average age of 76.0 ± 7.8 years old (from 65 to 87). For each human subject, the extraction is carried out on left and right members in order to analyze the impact of laterality on the architecture and mechanical behavior of the bone. Attention was paid to the preservation method to reduce bone drying before mechanical testing. Bodies are preserved using a body embalming procedure (distilled water, glycerine, methanol and phenol solution). For each bone, a 100 mm long diaphysis is extracted in the middle of the shaft. Macro mechanical samples were prepared and lathed after the extraction of a 50 mm parallelepipedic cortical sample using a surgical saw. The dog-bone shaped sample is obtained using a desktop lathe (Optimum TU 2004V). The targeted shape of the sample is to obtain a 6 and 2.5 mm effective length and diameter, respectively (Fig. 1). Between each step of the experimental procedure, the sample is kept hydrated in distilled water and stored at 4 °C. After μ-CT scan, samples were re-hydrated during at least 12 h at 4 °C. Before being mechanically tested, samples were acclimated at 21 °C in water for 1 h. Moreover, to reduce the impact of the preservation method on mechanical result, a maximum of 15 days was imposed between the extraction on the body and mechanical test. Thirty samples were obtained, scanned and subsequently tested. However, nine samples were inoperable for mechanical characterization.

Thirty samples were scanned along the longitudinal axis of the sample using a SkyScan 1172 high resolution desktop micro-CT scanner (80 kV, 100 μA, rotation step: 0.25° over 360°). The chosen voxel size is 2.94 μm which is enough to accurately detect vascular canals in cortical bone but not lacunar-canalicular network (which is not the aim of the current paper) [1]. Slices were reconstructed with NRecon 1.7.04 software. Due to micro-CT camera range, a few more than the middle part of the specimen is scanned (effective length and a part of radii, Fig. 1).

Then a python script is used to quantify three-dimensional geometrical features [12]. The novelty of the current procedure is the architectural study is able to consider the three-dimensional characteristics of the vascular network. Therefore, this

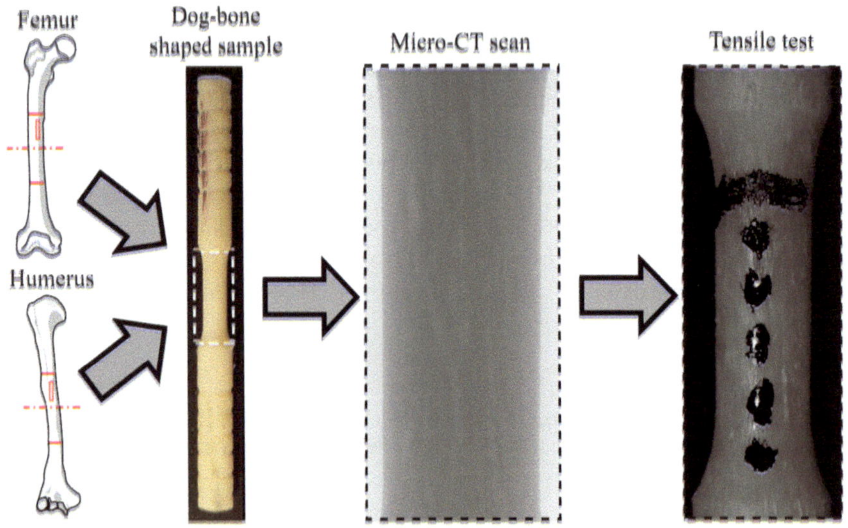

Fig. 1 Steps of the experimental procedure

Table 1 Definition of mechanical features computed from stress-strain curve

Mechanical features	Notation		Units
	Corrected	Apparent	
Elastic modulus	E_{corr}	E_{app}	(GPa)
Yield stress	$\sigma_{y,corr}$	$\sigma_{y,app}$	(MPa)
Linear toughness	$U_{l,corr}$	$U_{l,app}$	(mJ/mm^3)
Post-linear toughness	$U_{pl,corr}$	$U_{pl,app}$	(mJ/mm^3)
Total toughness	$U_{t,corr}$	$U_{t,app}$	(mJ/mm^3)
Ultimate stress	$\sigma_{u,corr}$	$\sigma_{u,app}$	(MPa)
Ultimate strain	ε_u		(mm/mm)

methodology uses the edges detection on every thresholded slices (Otsu threshold method [15]) to compute 14 and 3 features for each canal and connectivity, respectively. Canals are discretized by short segments which links two consecutive pores from different slices. Hence, the orientation, the real diameter, the length or aspect ratio of canals can be evaluated. In addition, connectivities are quantified by three features such as the opening angle which describes the angle between two connected canals.

Quasi-static tensile tests until failure were performed for each sample. An ElectroPuls E3000 (INSTRON) electromagnetic machine associated with a dynamic load capacity of 3 kN is used. Tests were conducted at a strain rate of 1.4×10^{-4} s^{-1}. Each test was carried out in a room where ambient temperature is kept at 21 °C. Due to the highly fragile mechanical behavior of the cortical bone, displacement measurement requires a high resolution camera. Hence, strains are measured using Manta G-895 camera (4000 × 2000 pixels) running at 1 Hz. A 2.8 μm pixel size was obtained. Right before testing, black dots are drawn along the effective length (Fig. 1). Images are post-processed using Digital Image Correlation and Tracking Matlab toolbox. Strain is defined as the mean displacement of all pairs of tracked points divided by the initial length (also called engineering strain). Different types of conventional stress are defined. Indeed, thanks to the μ-CT scan all along the effective length, Tissue Area (TA) and Bone Area (BA) can be computed. The first one defines the complete area of the response surface whereas the second one subtracts voids of the tissue area. Hence, apparent stress (σ_{app}) and corrected stress (σ_{corr}) are defined as the ratio between load and the minimum of TA and BA, respectively. As a consequence, each mechanical feature can be computed for both stresses and will be called in this paper with the corresponding previously defined suffix (Table 1). The objective of measuring two types of stress is to make data comparable with other studies where the type of response area isn't specified and to correlate mechanical data with architecture ones.

Figure 2 displays a typical stress-strain curve obtained from the tensile test until failure of a cortical bone sample. To compute elastic features, a linear fitting is applied to the curve on the elastic part which is supposed to be linear. Strain below 0.2% is commonly admitted as to be located in the elastic behavior of the bone [4, 5]. From this fitting, elastic modulus is computed as the rope of the linear fitting. The end of the linear part, which can be approximated as a macroscopic yield limit, is estimated as the point where deviation between curve and linear fitting exceeds 5%. Likewise, toughness can be evaluated from the area under the curve for both linear and post-linear part. Furthermore, total toughness is defined as the sum of the two previously computed values (see Table 1 for respective notation). Finally, ultimate stresses and strain are assessed.

Results and Discussions

A database has been implemented in Python 3.7 in order to easily sort, plot and compare statistical values. Statistical study was conducted using Numpy 1.15, Scipy 1.1.0 associated with Pandas 0.23.3 for storing data and Seaborn 0.9.0 Python libraries. From this database, 30 samples were extracted (14 femoral and 16 humeral samples) and valid for architectural study. Due to multiple reasons (slippage, crushing . . .), nine samples were removed from statistical assessment for mechanical features. Consequently, 21 samples (12 femoral and 9 humeral samples) are available for complete analysis (architecture and mechanical). Results are stored in Table 2 where mean values and associated standard deviation is displayed. Figure 3 shows stress-strain curves obtained for the entire set of tests.

Table 2 is quite revealing in several ways. First, for both types of elastic modulus, humeral samples show a stiffer elastic behavior than femoral ones. Second, assessed ultimate strains of specimens extracted from femurs are higher than humeri. However, no significant difference is revealed for ultimate stress. These comparative results are strengthened by curves on Fig. 3. This figure reveals that how important it is to keep subject information for each sample. Actually, the first printing of these curves seems to display highly dispersive results. Nevertheless, by individually comparing results from each subject, mechanical behavior appears to be closely linked with the subject, especially for femoral samples. All the subjects show a

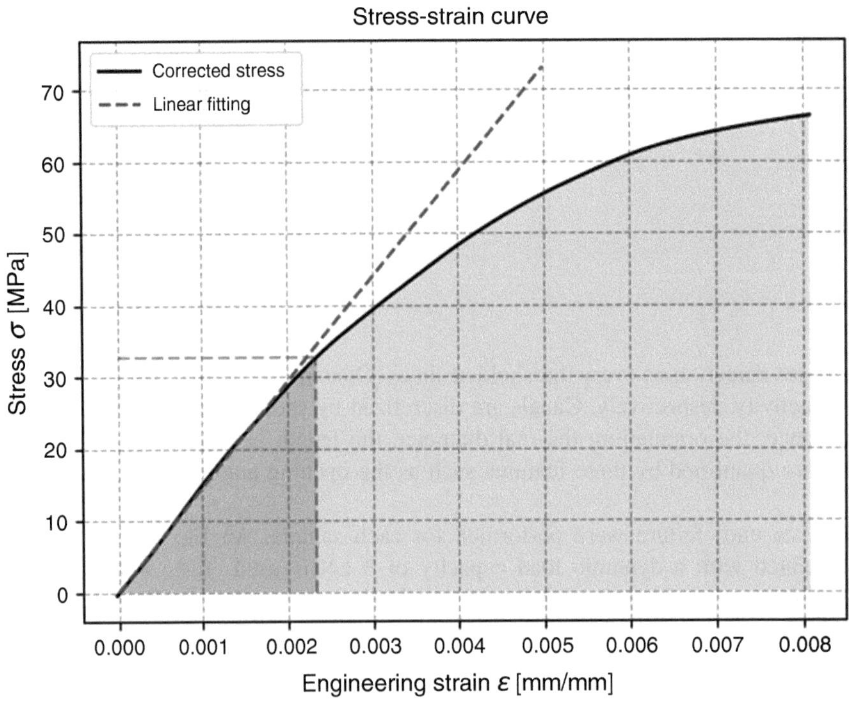

Fig. 2 Typical stress-strain curve. Linear dotted line represents the linear fitting to compute elastic modulus. Different grayscale areas show the linear toughness and post-linear toughness, respectively

Table 2 Mechanical features obtained from quasi-static tensile test of bearing and non-bearing bones

Mechanical features		Global (N = 21)	Femur (N = 12)	Humerus (N = 9)	Units
Elastic modulus	E_{corr}	17.2 ± 2.4	15.9 ± 2.0	19.0 ± 1.6	(GPa)
	E_{app}	15.9 ± 2.1	14.9 ± 1.8	17.3 ± 1.5	
Yield stress	$\sigma_{y,corr}$	49.0 ± 18.5	43.8 ± 17.5	55.8 ± 17.5	(MPa)
	$\sigma_{y,app}$	44.9 ± 15.22	41.1 ± 16.2	50.0 ± 11.8	
Linear toughness	$U_{l,corr}$	0.085 ± 0.062	0.077 ± 0.064	0.095 ± 0.057	(mJ/mm^3)
	$U_{l,app}$	0.077 ± 0.053	0.073 ± 0.060	0.083 ± 0.041	
Post-linear toughness	$U_{pl,corr}$	0.700 ± 0.464	0.860 ± 0.451	0.456 ± 0.389	(mJ/mm^3)
	$U_{pl,app}$	0.651 ± 0.428	0.803 ± 0.412	0.450 ± 0.360	
Total toughness	$U_{t,corr}$	0.785 ± 0.467	0.937 ± 0.465	0.581 ± 0.384	(mJ/mm^3)
	$U_{t,app}$	0.728 ± 0.432	0.875 ± 0.425	0.533 ± 0.358	
Ultimate stress	$\sigma_{u,corr}$	93.1 ± 18.5	92.6 ± 21.6	93.4 ± 13.2	(MPa)
	$\sigma_{u,app}$	86.1 ± 16.8	86.8 ± 19.6	85.2 ± 12.1	
Ultimate strain	ε_u	1.16 ± 0.45	1.36 ± 0.41	0.90 ± 0.35	(%)

stiffer elastic behavior and a weaker ultimate strain for humeral samples. Interestingly, failure arises at a similar stress for both of the bones (individual comparison). A possible explanation for this might be that the cortical bone matrix breaks at a constant stress (which is subject dependent) and the different architectures impact the ultimate strain. Thus, various mechanical behavior can be found. Concerning toughness, the total toughness mainly comes from the post-linear behavior and values are higher for the bearing bone than the non-bearing bone. It can be explained by the higher ultimate elongation of femur samples. These results are mainly consistent with Bry et al. study [8]. However, corrected and apparent moduli of femur are lower than those measured in studies [4, 5, 11, 16]. These differences could be explained by various reasons. For instance, preservation methods before and after specimen extraction can differ between several studies.

Moreover, some of these studies test bone with three-point bending tests. Using this method makes sample extraction easier but requires assumptions to compute elastic modulus (beam theory). In addition, during a bending test, specimen is tested in both tension and compression. It is commonly verified that compressive behavior of the cortical bone differs from tensile one [4, 11, 17]. The comparison of humeral data with bibliography is not trivial. As a matter of fact, humerus is poorly

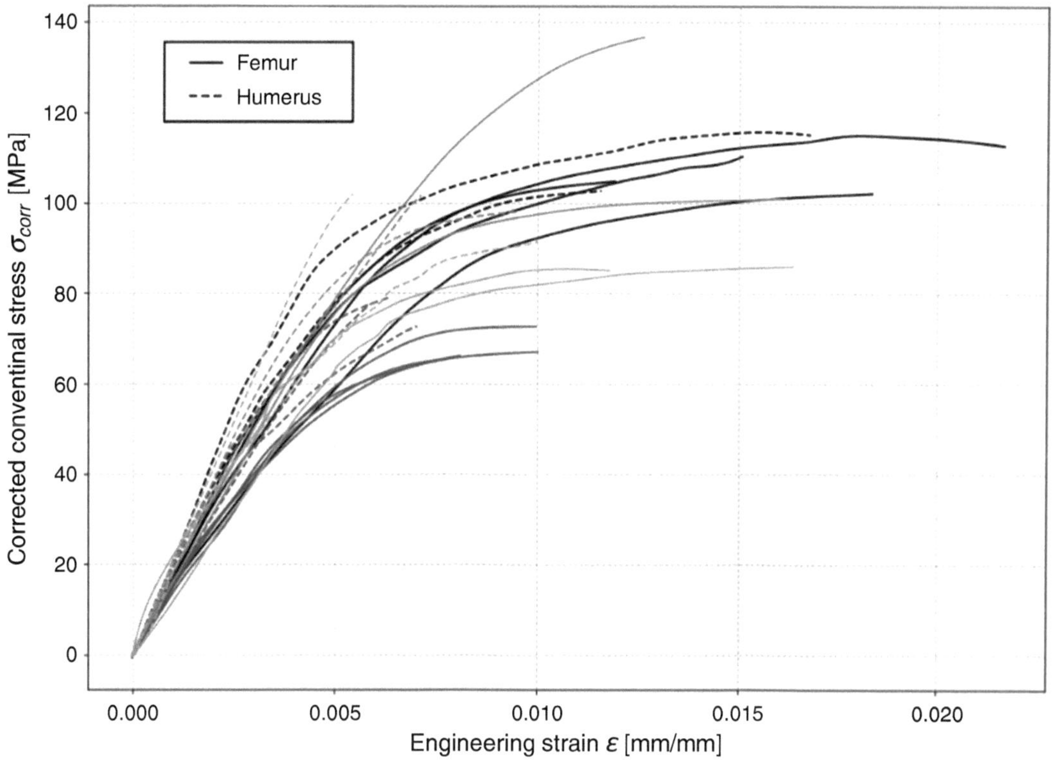

Fig. 3 Stress-strain curves for all valid test (N = 21). Femoral and humeral samples are highlighted in full and dotted lines, respectively. Different subjects are exhibited with different grayscale colors

Table 3 Mean values and standard deviation obtained from architectural assessment

Architectural data	Notation	Global (N = 30)	Femur (N = 14)	Humerus (N = 16)	Units
Volume porosity	BV/TV	6.39 ± 3.28	5.20 ± 1.26	7.44 ± 4.06	(%)
Canal density	Ca.D	75.40 ± 14.65	80.59 ± 18.25	70.85 ± 8.18	$(mm)^{-3}$
Connectivity density	Conn.D	23.62 ± 4.30	25.78 ± 3.65	21.73 ± 3.93	$(mm)^{-3}$
Canal real length	Ca.rLe	245.08 ± 29.42	254.79 ± 23.30	236.58 ± 31.52	(μm)
Canal real diameter	Ca.rDm	74.15 ± 21.71	67.31 ± 12.01	80.14 ± 26.10	(μm)
Canal longitudinal angle	Ca.x	27.70 ± 1.83	26.60 ± 1.59	28.66 ± 1.44	(deg)
Canal transverse angle	Ca.y	0.89 ± 5.30	0.83 ± 4.10	0.93 ± 6.15	(deg)
Canal 3D aspect ratio	Ca.AR3D	4.78 ± 1.04	5.19 ± 1.00	4.41 ± 0.94	(–)
Connectivity opening angle	Conn.α	34.94 ± 3.02	34.02 ± 2.32	35.75 ± 3.32	(deg)

Notation formalism is the same as the one used and described in [12]

investigated. One of the reasons for this lack could be that the very thin cortical shell of a human humerus makes extraction highly complex compared with the femur.

The architectural results obtained thanks to the python script are displayed in Table 3. Bar plot of the porosity for each laterality of each bone is shown in Fig. 4.

The assessment of the architecture reveals some differences between bearing and non-bearing bones. Indeed, the porosity (also called BV/TV) seems to be higher and more dispersive in humerus diaphysis than in the femur. This result is comforted by Fig. 4 where left and right members are exhibited. On this figure, porosity of the left humerus is higher than the right side whereas femur shows small differences. This trend is verified by individually comparing subject (which is not shown in this paper). No comparative work was found. However, Perchalski et al. study [9] proves that there is a development asymmetry between left and right humeri throughout the life. The authors explain that a dominant arm creates this asymmetry during development. Results from the current study seek to extend this conclusion to a microscopic level in the humeral cortical shell. In the architecture, the differences of porosity can be explained by different sizes of canals and density. In fact, the canal length seems to be constant whatever the bone. However, the canal diameter positively affects the porosity for both

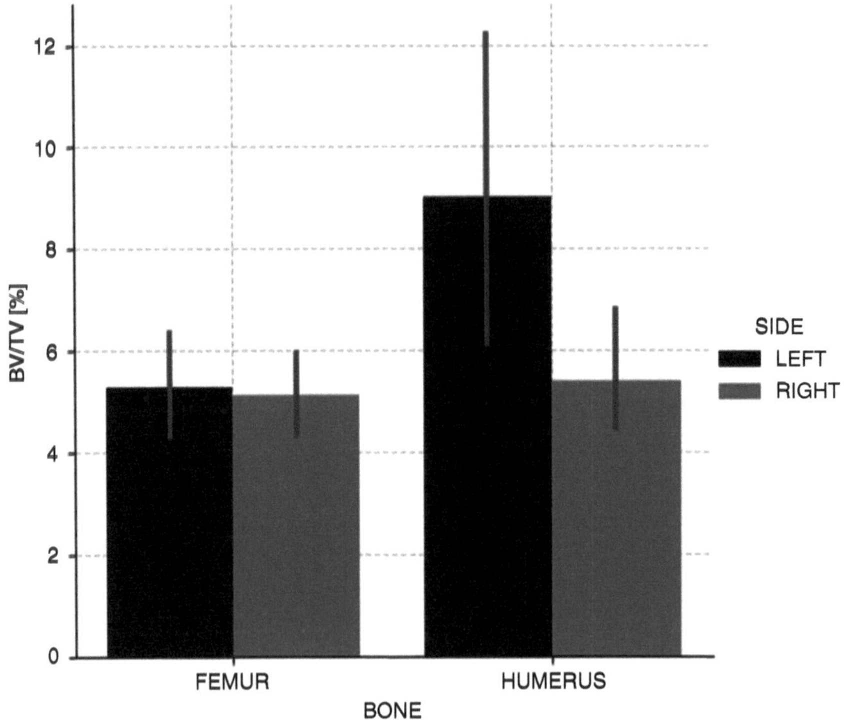

Fig. 4 BV/TV bar plot by highlighting the side member

bones. The rest of the architecture features (global orientation, connectivity opening angle) appear to be bone independent. Yet no comparative studies were found for human bone, and consequently these values may be useful for further studies.

Architecture differences between the two bones could partially explain the differences in the macroscopic mechanical behavior. In fact, a correlation study (Pearson and Spearman correlation) was performed so as to investigate a potential relation between two features. Figure 5 shows the scatter plot of ultimate strain versus the mean opening angle of connectivity. Figure 6 displays the volume porosity versus the mean value of canal diameters.

The Spearman correlation value obtained from the relation shown in Fig. 5 suggests a negative relation between the two features (−0.56) whatever the bone. However correlation value is more important for humeral (−0.67) than femoral (−0.48) samples. Obviously, results look like to be dispersive, but they suggest that there is an existing impact of the vascular network on the macro-mechanical behavior. This relation means that less connectivity is tightened more the failure strain is important. Moreover, this conclusion depends on the bone. This trend can explain the differences between femur and humerus concerning the ultimate elongation. Another interesting correlation was found between the apparent corrected modulus E_{app} and BV/TV. However, due to the small number of samples ($N = 21$), correlation work needs a more substantial database in order to strengthen some current trends and be more representative.

Due to the impact of the volume porosity on the mechanical behavior, it would be interesting to investigate the reasons of differences of this architectural feature. Figure 6 and Spearman correlation value (0.94) helps to conclude that BV/TV highly depends on the canal diameter. No other significant correlations were found. This trend suggests that porosity depends on the geometrical features of the vascular canal network, and subsequently depends on the bone remodeling activity. This conclusion needs to be further assessed with a higher number of human subjects so as to obtain various vascular network arising from various remodeling activities.

Conclusion

The purpose of the current work is to show the different mechanical behaviors between a bearing and a non-bearing human bone. Therefore, these differences are investigated by assessing micro-architecture using tomographic data. The study is focused on left and right humeri and femurs, located in the upper and lower member, respectively. Samples were dumbbell shaped in order to be both fully scanned on the effective length and mechanically tensile tested. Hence, two stresses can

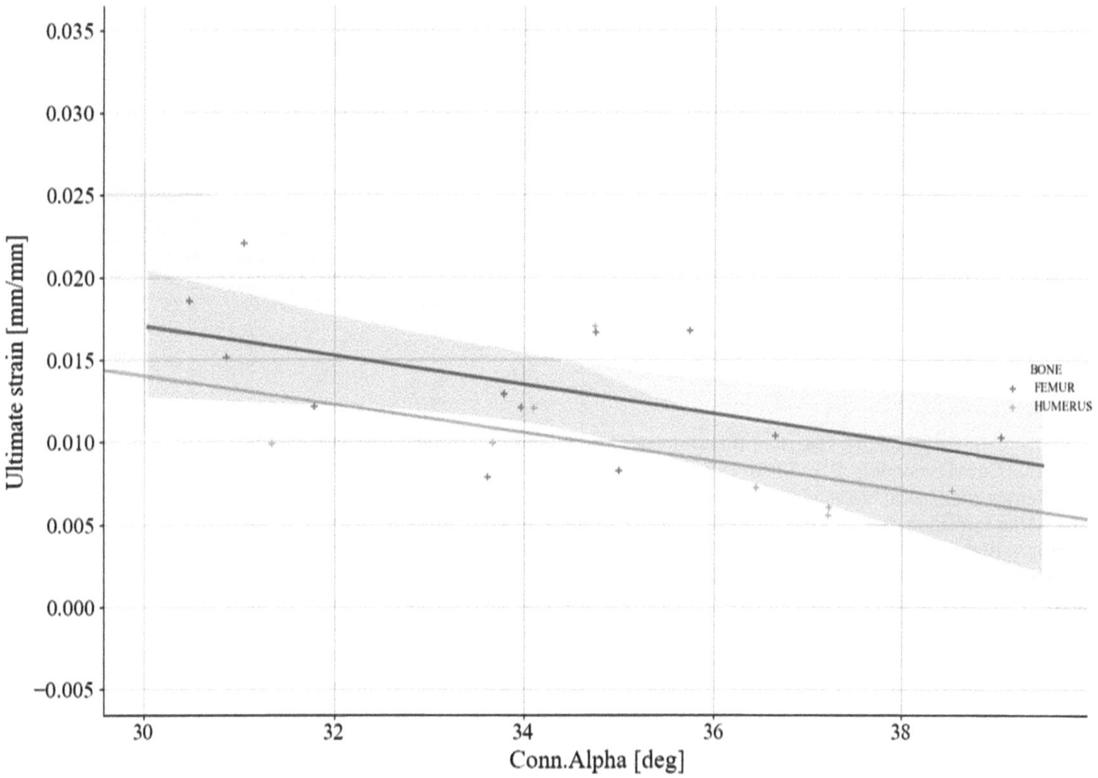

Fig. 5 Ultimate strain versus opening angle of connectivity (N = 20). An outlier humeral sample was removed due to its high Conn.α value (46°). Straight lines show the equivalent linear regression and colored area the associated 95% confidence interval

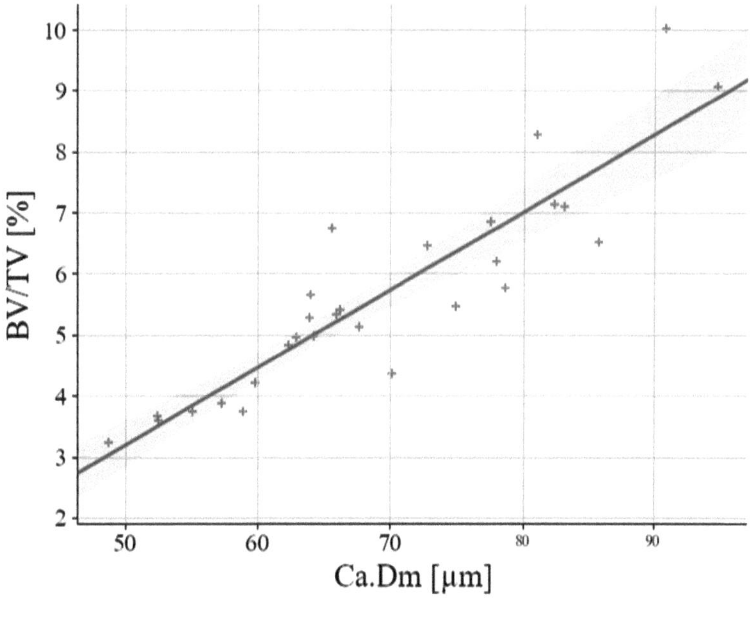

Fig. 6 Scatter point of volume porosity versus mean value of canal diameter. The plot was obtained from all samples available for architectural analysis. However, two outliers were removed due to their non-representative BV/TV values (over 16%) (N = 28). Straight line show the equivalent linear regression and colored area the associated 95% confidence interval

be computed from two types of response surface obtained from tomography. The novelty of this survey is the use of a Python script which spatially described the cortical architecture and investigate the impact on the tensile results. Moreover, to reduce bone drying, all the samples were tested 15 days maximum after harvest on the human body. Four male subjects were studied. Differences between humerus and femur are found. First, elastic modulus and ultimate strain are different.

Second, µ-CT scans reveal that volume porosity and canal diameter are higher in humerus than femur. Similarly, important disparities are found in humerus which suggests that there is a significant effect of a dominant arm on the architecture. Finally, some correlations linked mechanical features with vascular canals characteristics such as the angle of connectivity with the ultimate strain. Nonetheless, these conclusions need to be deeply investigated with a more substantial database of human subjects.

Acknowledgements This research is funded by the French Ministry of Higher Education and Research and is carried out within the framework of the CNRS Research Federation on Ground Transports and Mobility, in articulation with the ELSAT2020 project supported by the European Community, the Hauts de France Regional Council. The authors gratefully acknowledge the support of these institutions.

References

1. D.M.L. Cooper, C.E. Kawalilak, K. Harrison, B.D. Johnston, J.D. Johnston, Cortical bone porosity: what is it, why is it important, and how can we detect it? Curr. Osteoporos. Rep. **14**(5), 187–198 (2016)
2. H.M. Frost, Tetracycline-based histological analysis of bone remodeling. Calcif. Tissue Res. **3**(3), 211–237 (1969)
3. H.M. Frost, Mean formation time of human osteons. Can. J. Biochem. Physiol. **41**(5), 1307–1310 (1963)
4. M.J. Mirzaali, A. Bürki, J. Schwiedrzik, P.K. Zysset, U. Wolfram, Continuum damage interactions between tension and compression in osteonal bone. J. Mech. Behav. Biomed. Mater. **49**, 355–369 (2015)
5. U. Stefan, B. Michael, S. Werner, Effects of three different preservation methods on the mechanical properties of human and bovine cortical bone. Bone **47**(6), 1048–1053 (2010)
6. X. Neil Dong, X. Edward Guo, The dependence of transversely isotropic elasticity of human femoral cortical bone on porosity. J. Biomech. **37**(8), 1281–1287 (2004)
7. F. Libonati, L. Vergani, Bone toughness and crack propagation: an experimental study. Procedia Eng.**74**, 464–467 (2014)
8. R. Bry, *Contribution à l'étude de la variabilité des propriétés mécaniques de l'os cortical diaphysaire d'un os porteur (fémur) et non-porteur (humérus)*. PhD thesis. (Valenciennes, 2015). http://www.theses.fr/2015VALE0022
9. B. Perchalski et al., Asymmetry in the cortical and trabecular bone of the human humerus during development. Anat. Rec. (Hoboken) **301**(6), 1012–1025 (2018)
10. F. Vandenbulcke et al., *On the mechanical characterization of human humerus using multi-scale continuum finite element model*, in 2012 IRCOBI Conference, Dublin, 2012, pp. 598–610
11. M.J. Mirzaali et al., Mechanical properties of cortical bone and their relationships with age, gender, composition and microindentation properties in the elderly. Bone **93**(Supplement C), 196–211 (2016)
12. X. Roothaer, R. Delille, H. Morvan, B. Bennani, E. Markiewicz, C. Fontaine, A three-dimensional geometric quantification of human cortical canals using an innovative method with micro-computed tomographic data. Micron **114**, 62–71 (2018)
13. X. Roothaer, R. Delille, H. Morvan, B. Bennani, E. Markiewicz, C. Fontaine, « *Quantitative method for the three-dimensional assessment of human cortical long-bone architecture based on micro-CT images* ». CMBBE 2018. Lisbon, Portugal (2018)
14. D.M.L. Cooper, A.L. Turinsky, C.W. Sensen, B. Hallgrímsson, Quantitative 3D analysis of the canal network in cortical bone by micro-computed tomography. Anat. Rec. B. New Anat. **274B**(1), 169–179 (2003)
15. N. Otsu, A threshold selection method from gray-level histograms. IEEE Trans. Syst. Man Cybern. **9**(1), 62–66 (1979)
16. G. Zhang, X. Deng, F. Guan, Z. Bai, L. Cao, H. Mao, The effect of storage time in saline solution on the material properties of cortical bone tissue. Clin. Biomech. **57**, 56–66 (2018)
17. L. Duchemin et al., Prediction of mechanical properties of cortical bone by quantitative computed tomography. Med. Eng. Phys. **30**(3), 321–328 (2008)

3D High-Speed Digital Image Correlation (3D-HSDIC) to Study Damage of Human Eardrum Under High-Pressure Loading

Payam Razavi, Haimi Tang, Koohyar Pooladvand, Cassia Larson, Eli W. Frank, John J. Perkoski, Jacquelyn Y. Roberge, Jessica C. Walsh, John J. Rosowski, Jeffrey T. Cheng, and Cosme Furlong

Abstract High-intensity impulsive sounds (e.g., explosions, firearms, etc.) are known to damage the human eardrum (Tympanic Membrane, TM), and produce conductive hearing loss. However, the eardrum failure mechanism under high-pressure loading has not been well studied. Our previous investigations have shown the feasibility of using full-field-of-view 3D High-Speed Digital Image Correlation (3D-HSDIC) with a custom-made loading apparatus to study acoustically induced damage on artificial membranes. In this paper, we present new measurements on multiple human post-mortem eardrums. The dynamic 3D displacements and transient shape changes of the entire eardrum surface during the rupture are simultaneously quantified at high frame rates (i.e., >100,000 Hz) using the 3D-HSDIC method over the rapid time-course of the TM response. The results describe the high strain-rate and large displacements of the eardrum from the initial stages of rapid pressurization up to complete eardrum failure. The high spatio-temporal resolution measurements allow the determination of eardrum mechanical properties under high-pressure loading. This study indicates the potential utility of high-speed DIC to study high pressure induced TM failure mechanisms, which has impact on developing new hearing protection devices. Future measurements will be performed with a miniaturized optical system and an updated high-pressure loading apparatus designed using advanced thermo-acousto-fluidic numerical modeling and high-speed Schlieren imaging techniques.

Keywords 3D high-speed digital image correlation (3D-HSDIC) · Acoustic trauma · Blast loading apparatus · Acoustic to solid interaction · Hearing protection · Tympanic membrane

Introduction

High-intensity impulsive sounds caused by explosions, large caliber military ammunition, etc., can damage the human eardrum (tympanic membrane, TM), and cause mild to severe conductive hearing loss. Twenty-eight percent of all military personnel experience some degree of hearing loss post-deployment [1, 2]. Therefore, designing better ear protection devices for personnel exposed to high amplitude sounds will help prevent the TM from rupturing and will reduce hearing loss risks.

P. Razavi (✉) · H. Tang · K. Pooladvand · E. W. Frank · J. J. Perkoski · J. Y. Roberge · J. C. Walsh
Center for Holographic Studies and Laser micro-mechaTronics (CHSLT), Worcester, MA, USA

Mechanical Engineering Department, Worcester Polytechnic Institute, Worcester, MA, USA
e-mail: prazavi@wpi.edu; htang3@wpi.edu; kpooladvand@wpi.edu; ewfrank@wpi.edu; jjperkoski@wpi.edu; jyroberge@wpi.edu; jcwalsh@wpi.edu

C. Larson
School of Engineering and Applied Science, Harvard University, Boston, MA, USA
e-mail: c_larson@college.harvard.edu

J. J. Rosowski · J. T. Cheng
Eaton-Peabody Laboratory, Massachusetts Eye and Ear Infirmary, Boston, MA, USA

Department of Otolaryngology–Head and Neck Surgery, Harvard Medical School, Boston, MA, USA
e-mail: John_Rosowski@meei.harvard.edu; Tao_Cheng@meei.harvard.edu

C. Furlong
Center for Holographic Studies and Laser micro-mechaTronics (CHSLT), Worcester, MA, USA

Mechanical Engineering Department, Worcester Polytechnic Institute, Worcester, MA, USA

Eaton-Peabody Laboratory, Massachusetts Eye and Ear Infirmary, Boston, MA, USA

Department of Otolaryngology–Head and Neck Surgery, Harvard Medical School, Boston, MA, USA
e-mail: cfurlong@wpi.edu

M. E. Grady (ed.), *Mechanics of Biological Systems and Materials & Micro-and Nanomechanics, Volume 4*,
Conference Proceedings of the Society for Experimental Mechanics Series, https://doi.org/10.1007/978-3-030-30013-5_10

To achieve this goal, better models are needed to predict blast injuries, especially to understand the dynamics of the TM rupture process. Here we describe, a high-speed, non-intrusive full-field-of-view metrology that we developed to study TM rupture damage due to impulsive high amplitude excitations.

The objective of high-speed full-field-of-view measurements is to provide the following information to complement the existing models: (1) shape variations of the membrane due to large deformation under high pressure load affect the stress concentration which may relate to the location of the rupture initialization; (2) data describing temporal and spatial variations in displacements and strains during the rupture will provide valuable information for model verification purposes; (3) the direction and amplitude of the velocities during the rupture initialization and propagation are related to the characterization of the blasts and the orthotropic mechanical properties of the membrane and their sound-matter interaction.

Methods

The Digital Image Correlation (DIC) system used in this study consisted of two high-speed (up to 2.1 M fps) cameras in a stereo configuration with an angle of 30° that can perform 3D measurements at micrometer spatial resolution (shown in Fig. 1). Three LED lamps are used to provide short-exposure illumination. A rapid step change in static air pressure inside the middle ear cavity is used to rupture the TM. High-frequency pressure sensors and microphones are deployed to measure the air pressure and sound on both sides of the TM. The release of the air into the middle ear and the trigger of the cameras are synchronized via computer-controlled data acquisition.

Human temporal bones were prepared by removing the boney and cartilaginous ear canals to allow a full view of the TM surface. The TMs were painted with white and back speckles (Fig. 2) to provided trackable features for the DIC algorithm. The middle ear was opened via the facial recess to permit pressurization of the middle ear cavity and pressure measurement. The recess opening and the bony surfaces around the ear were sealed with dental cement and silastic impression material to prevent air leakage.

To quantify the rupturing process the von Mises strains ε_e is used which is defined by Eq. (1) [3],

$$\varepsilon_e = \sqrt{\frac{1}{2}\left[(\varepsilon_1 - \varepsilon_2)^2 + (\varepsilon_2 - \varepsilon_3)^2 + (\varepsilon_1 - \varepsilon_3)^2\right]}, \tag{1}$$

where, ε_i denotes the principal strains.

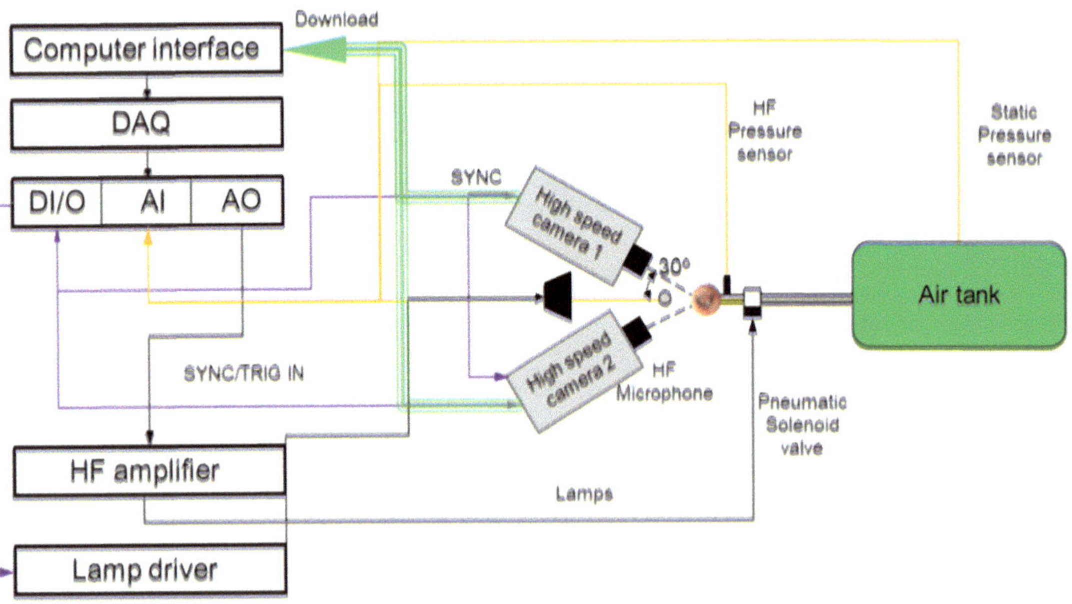

Fig. 1 DIC configuration and experiment setup. Two high speed cameras running at 100,000 fps are synchronized with the release of air into the middle ear to capture the rupture of the tympanic membrane

Fig. 2 Spackles pattern on the tympanic membrane. A Master airbrush (Model S68) is used with HOLBEIN water-soluble oil color (TITANIUM WHITE DU 461) to paint the surface white and a modified Master airbrush (Model G222) is used with HOLBEIN water-soluble oil color (SPINEL BLACK DU 352) to speckle the surface

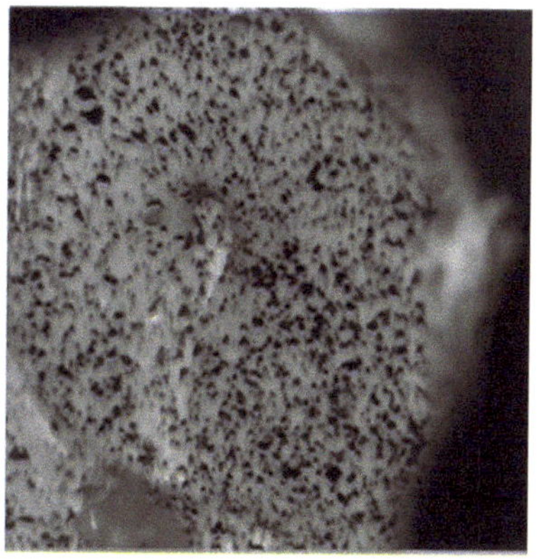

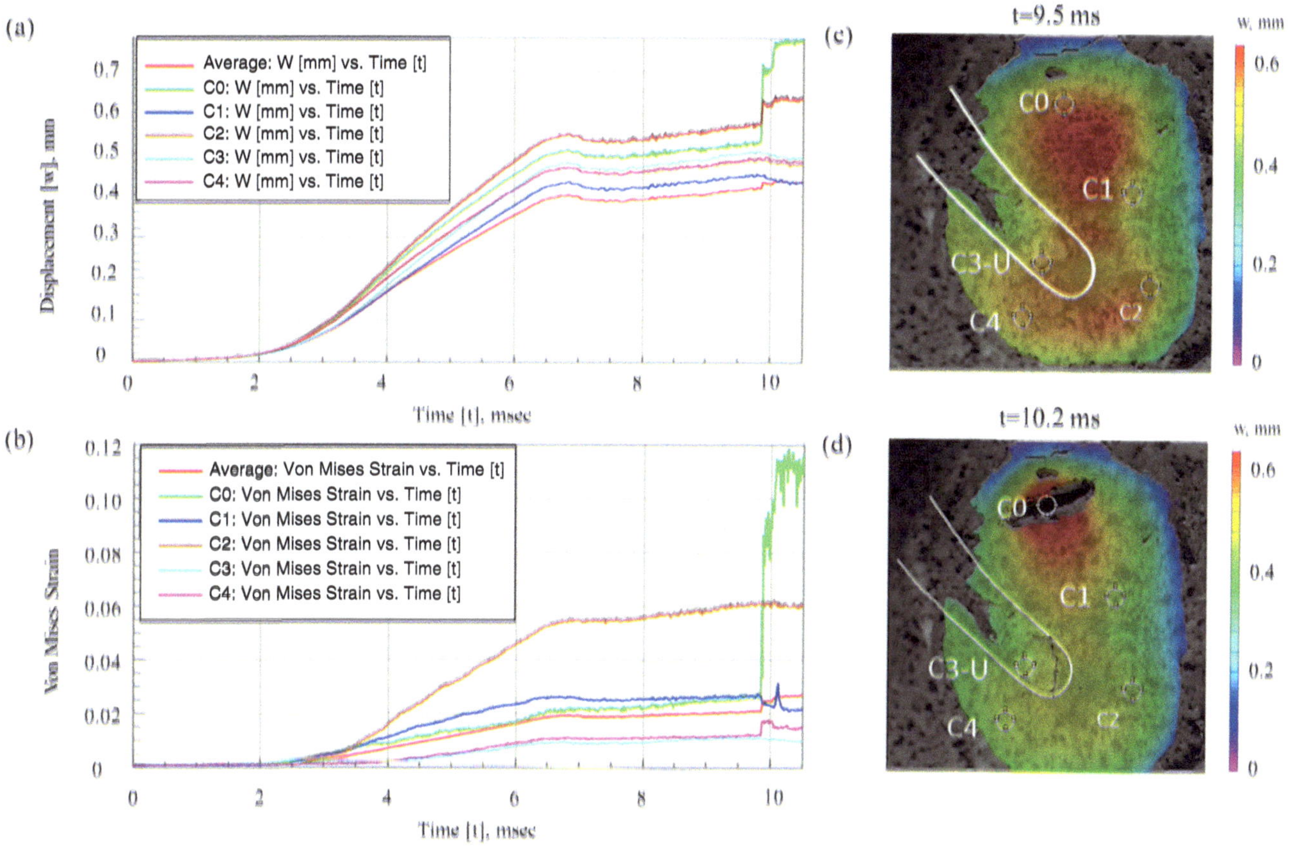

Fig. 3 DIC results from TM1: (**a**) out-of-plane displacement of six points on the TM surface starting from the initiation of pressurization (time 0). (**b**) the temporally varying von Mises strain computed from the deformations. (**c**) The displacements before the rupture, t = 9.5 ms. (**d**) the displacements after the rupture, t = 10.20 ms. The rupture started with a crack at location C0. Note both the displacement and von Mises strain at C0 location (green curves in (**a**) and (**b**)) show a rapid increase at ~9.5 ms when the crack first occurs

For the current configuration, principal plane strain condition is assumed ($\varepsilon_3 = 0$). Therefore, the von Mises ε_e strains is simplified as Eq. (2):

$$\varepsilon_e = \sqrt{{\varepsilon_1}^2 - \varepsilon_1\varepsilon_2 + {\varepsilon_2}^2}, \tag{2}$$

Results

Three TMs were successfully ruptured by rapid pressurization (within 10 ms). The out-of-plane displacement, w, and von Mises strain, ε_e are illustrated at selected points of each TM on Figs. 3, 4 and 5.

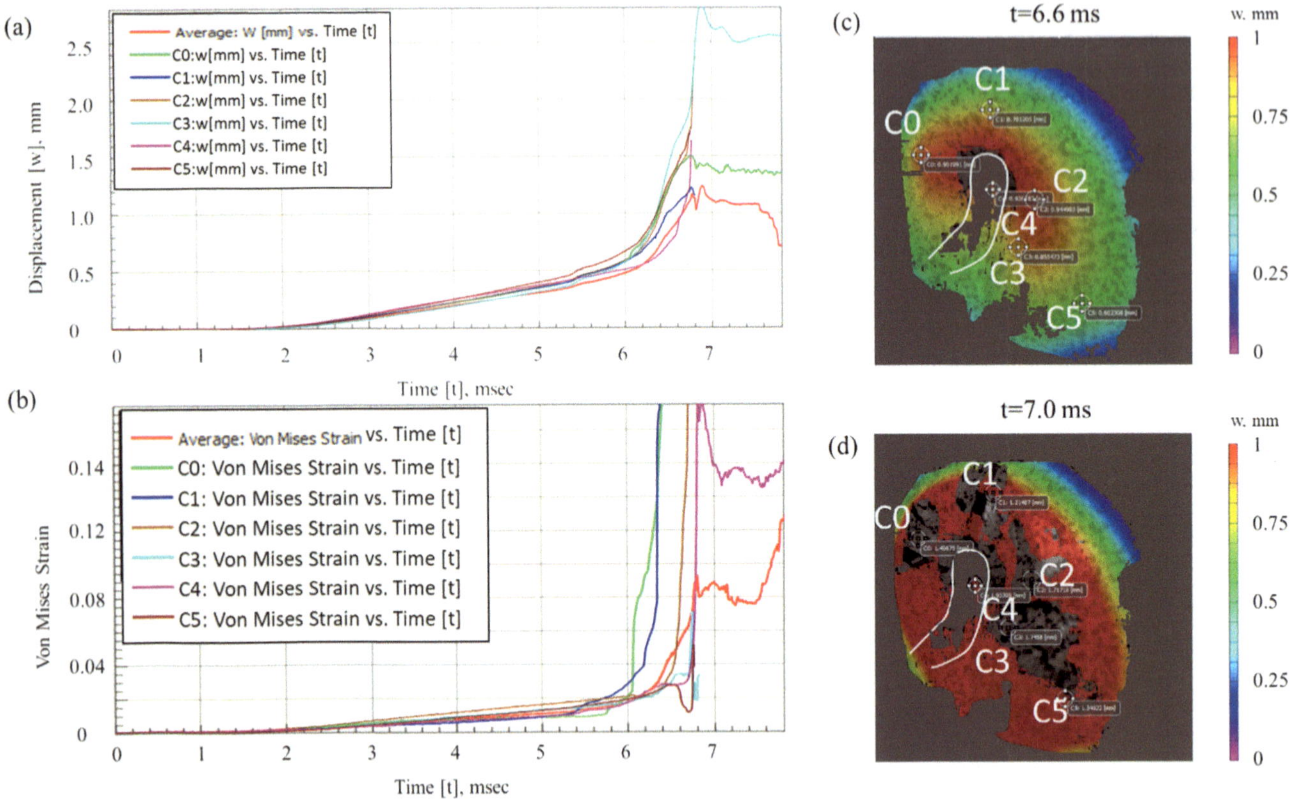

Fig. 4 DIC results from TM2: (**a**) out-of-plane displacement of six points on the TM surface starting from the initiation of pressurization (time 0). (**b**) the temporally varying von Mises strain computed from the deformations. (**c**) The displacements before the rupture, t = 6.6 ms. (**d**) the displacements after the rupture, t = 7.0 ms. Five cracks (C0, C1, C2, C4, C5) were observed during the rupture process

Based on the observation of out-of-plane displacement and von Mises strain, the TM failure process can be described in three sequential stages: (1) global TM displacement; (2) local strain concentrations and (3) local rupture and crack propagation mostly at the locations of high local strains.

To understand the mechanical properties of the TM under low to high loading levels, we plotted pressure vs. averaged surface displacement during the measurements (Fig. 6a) and compared it with the stress-strain constitutive behavior of several typical hyperplastic materials (Fig. 6b). Our preliminary observations of the mechanical properties of the TM show non-linear material characteristics and hardening which bear a similarity to hyperplastic constitutive materials under uniaxial stresses. These behaviors are reported for biological tissues such as epidermis in skin [5]. Our methodology can provide more details and a plethora of temporal-spatial data concerning material properties and failure mechanism compared to indentation studies and can lead to improved estimates of the properties of anisotropic biological tissues.

The crack opening speeds and opening directions of all the ruptures from TM1 to TM 3 are calculated by detecting the edge of the crack during rupture as shown in Fig. 7. The perforation in TM1 and TM3 produced a single circumferentially located crack that increased in length at a rate of about 100 mm/ms. While in TM2 all 5 cracks grew at slower speeds in a radial direction. We suspect that there are air-leaks in the TM1 and TM3 samples which reduced the rate of the pressure build up inside the middle ear cavity and led to different rupture profiles.

Conclusions and Future Work

We measured the full-field-of-view rupture of TMs by a high-speed DIC system and observed various crack opening speeds in circumferential and radial directions. Cracks along circumferential directions tend to open faster than the one along radial direction. The preliminary results establish the potential of high-speed DIC as a research tool to extend our understanding of tympanic membrane mechanical properties and rupture mechanism under blasts. In future work we will add thickness measuring capability (e.g. by OCT) to the current system to calculate stress-strain curve during the rupture process to allow

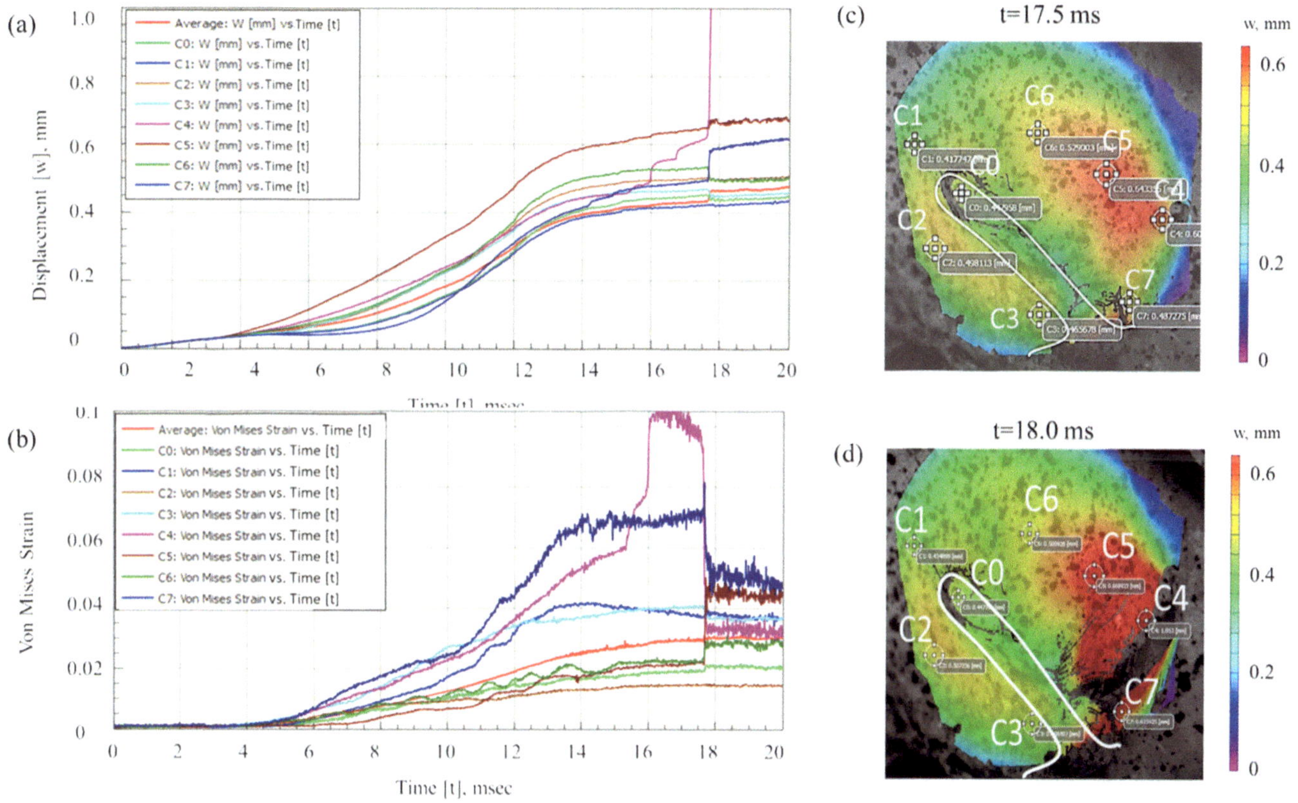

Fig. 5 DIC results from TM3: (**a**) out-of-plane displacement of six points on the TM surface starting from the initiation of pressurization (time 0). (**b**) the temporally varying von Mises strain computed from the deformations. (**c**) a displacements map at 17.5 ms right before the rupture. Prior to the main rupture high stress concentration and a small perforation was observed in location C4; (**d**) the displacements after the rupture, t = 18.0 ms.

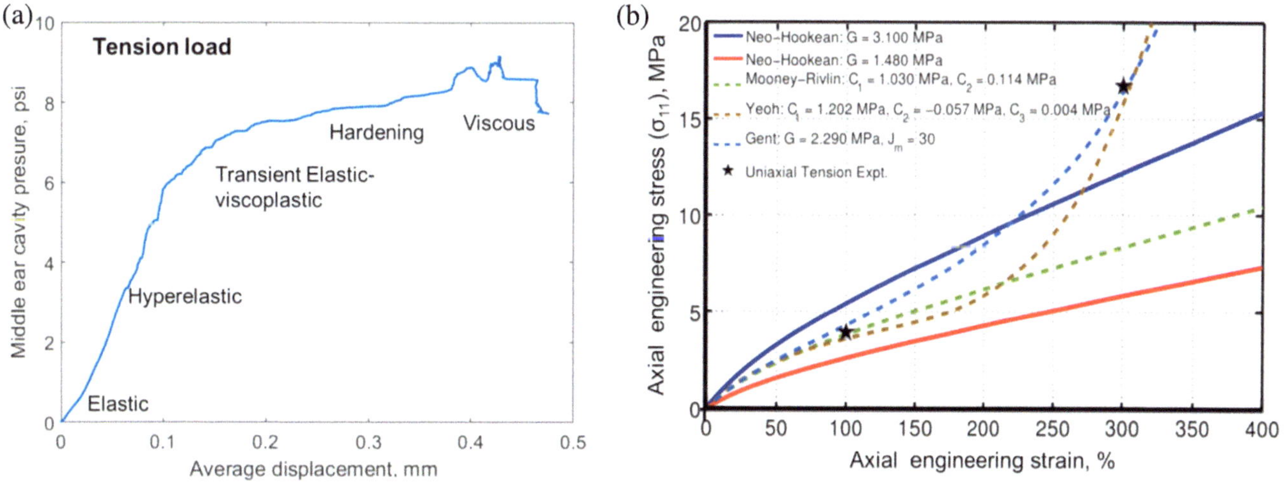

Fig. 6 Representative results of TM 3. (**a**) Pressure vs. displacement function during the rupture process suggests the TMs have typical elastic-viscoplastic-viscous behavior under different loading conditions; (**b**) Typical uniaxial stress–strain responses for different hyperplastic material models such as Neo-Hookean, Mooney-Rivlin, Yeoh, and Gent deformed at constant strain rate [4]

the analysis of mechanical properties, and adapt the high-speed Schlieren imaging technique into the system to visualize sound-matter interactions between the TM and acoustic waves.

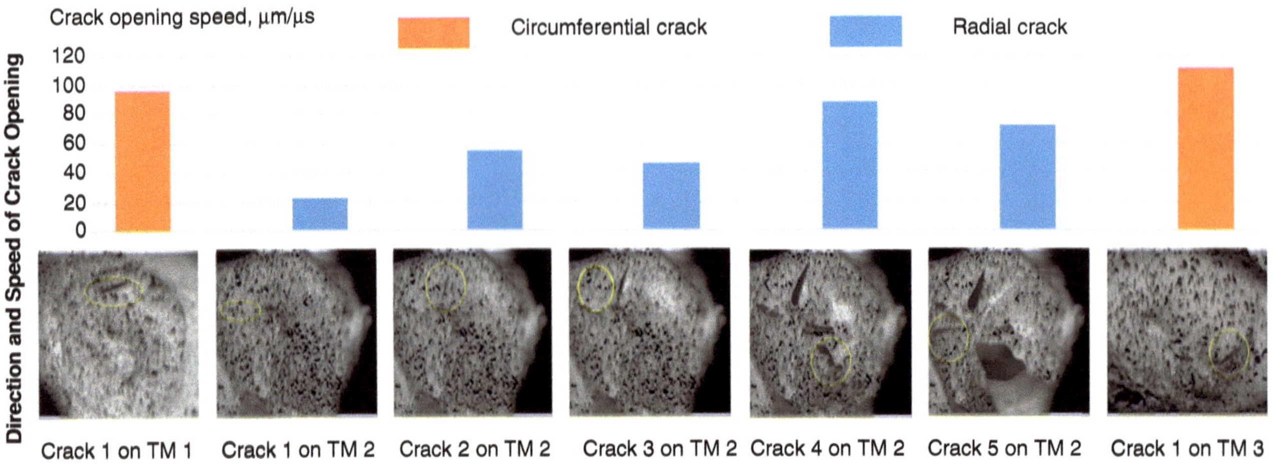

Fig. 7 Crack opening velocity and directions of TM1 to TM3. The upper panel shows the opening speed and directions are marked by different colors. The bottom panel shows the frame captured right after the crack with the crack location marked by yellow circle

References

1. Stanford Medicine, Long-term hearing loss from loud blasts may be treatable, researchers say, https://med.stanford.edu/news/all-news/2013/07/long-term-hearing-loss-from-loud-blasts-may-be-treatable-researchers-say.html. Accessed 1 Feb 2018
2. R. Gan, D. Nakmali, X. Ji, K. Leckness, Z. Yokell, Mechanical damage of tympanic membrane in relation to impulse pressure waveform – A study in chinchillas. Hear. Res. **340**, 25–34 (2016)
3. H.A. Bruck, S.R. McNeill, M.A. Sutton, W.H. Peters III, Digital image correlation using Newton-Raphson method of partial differential correction. Exp. Mech. **29**(3), 261–267 (1989)
4. Wikipedia, Hyperelastic material, https://en.wikipedia.org/wiki/Hyperelastic_material. Accessed 1 Feb 2018
5. A. Ní Annaidh et al., Characterization of the anisotropic mechanical properties of excised human skin. J. Mech. Behav. Biomed. Mater. **5**(1), 139–148 (2012)

Comparative Modal Analysis of the Tympanic Membrane Mechanics Between Normal and Experimentally Simulated Pathological Ears

Haimi Tang, Payam Razavi, Nima Maftoon, John J. Rosowski, Cosme Furlong, and Jeffrey T. Cheng

Abstract We are developing a High-speed Digital Holographic (HDH) system to measure acoustically induced transient displacements and shapes of live mammalian Tympanic Membrane (TM) for research and clinical applications. Currently, the HDH system measures the shape of the entire TM with a resolution of about 120 μm, and sound-induced displacements with magnitude resolutions of <10 nm and temporal resolutions of <20 μs in full-field of view. In this paper, we apply Experimental Modal Analysis (EMA) to the HDH results to extract modal parameters of the TM and quantitatively compare these parameters among different TMs of normal and pathological middle ears. Transient displacements in response to broadband acoustic clicks and the shape of cadaveric human TMs are measured before and after different simulated middle-ear pathologies including various levels of fluid in the middle ear cavity, stapes fixation, and incudo-stapedial (IS) joint interruption. The transient displacement of the TM along the sensitivity vector is supplemented with the 3-D TM shape information to derive the true displacement of the TM locally normal to the TM surface. The displacement is then used in an EMA framework to determine natural frequencies, damping and mode shapes of the TMs under the normal and different pathological middle-ear conditions.

Preliminary results show that the damping ratio of the TM at each natural frequency in the normal ear decreases as frequency increases. We also see differences in identified modal shapes and natural frequencies before and after various manipulations. We plan to identify trends in the data associated with different pathologies as well as test the sensitivity and selectivity of these analyses for clinical diagnosis. Due to the large amount of the data obtained from the HDH, Artificial Intelligence (AI) and Data Mining methods will be used to automate the EMA process and assist in the separation of normal and diseased states.

Keywords Artificial intelligence (AI) · Experimental modal analysis · High-speed digital holography · Human tympanic membrane · Middle-ear pathologies

---

H. Tang (✉) · P. Razavi
Center for Holographic Studies and Laser micro-mechaTronics (CHSLT), Worcester, MA, USA

Mechanical Engineering Department, Worcester Polytechnic Institute, Worcester, MA, USA
e-mail: htang3@wpi.edu; prazavi@wpi.edu

N. Maftoon
Systems Design Engineering Department, University of Waterloo, Waterloo, ON, Canada
e-mail: nmaftoon@uwaterloo.ca

J. J. Rosowski · J. T. Cheng
Eaton-Peabody Laboratory, Massachusetts Eye and Ear Infirmary, Boston, MA, USA

Department of Otology and Laryngology, Harvard Medical School, Boston, MA, USA
e-mail: John_Rosowski@meei.harvard.edu; Tao_Cheng@meei.harvard.edu

C. Furlong
Center for Holographic Studies and Laser micro-mechaTronics (CHSLT), Worcester, MA, USA

Mechanical Engineering Department, Worcester Polytechnic Institute, Worcester, MA, USA

Eaton-Peabody Laboratory, Massachusetts Eye and Ear Infirmary, Boston, MA, USA

Department of Otology and Laryngology, Harvard Medical School, Boston, MA, USA
e-mail: cfurlong@wpi.edu

M. E. Grady (ed.), *Mechanics of Biological Systems and Materials & Micro-and Nanomechanics, Volume 4*,
Conference Proceedings of the Society for Experimental Mechanics Series, https://doi.org/10.1007/978-3-030-30013-5_11

Introduction

The human eardrum or Tympanic membrane (TM) separates the outer ear from the middle ear and initiates the sound energy transmission to the middle and inner ear. The complex energy transformations start with TM motions that are affected by its complex geometry (shape and thickness), spatially varied mechanical properties and microstructure (radial and circumferential fibers), and the load from the ossicular chain and inner ear [1–3]. To better understand the acoustic to solid interaction of the TM in the sound transmission process, we developed High-speed Digital Holographic (HDH) system to quantify acoustically induced transient displacements and shapes of live mammalian TM.

The HDH system is currently capable [4] of: (1) shape determination of the entire TM with a spatial resolution of 120 μm; (2) measurements of nanometer scale acoustic click-induced transient displacements with a temporal resolution of less than 15 μs (i.e., >67 kHz); (3) adjustable shape measurement resolutions to account for complex TM shape depth variations; (4) adaptability to endoscopic measurements with small triangulation volumes. The HDH system's full-field of view measurement capability and the high spatial and temporal resolutions of the TM shape and broad-band transient displacement measurements allow us to apply Experimental Modal Analysis (EMA) based on the Frequency Response Function (FRF) matrix. Complex Mode Indicator (CMIF) functions are used as a condensate version of FRF to visualize and compare the FRFs.

Modal parameters including the modal shapes, displacement amplitudes, damping, and natural frequencies of complex motion modes are extracted quantitatively compare vibrations of different TMs and investigate the utility of this method to diagnose middle ear pathologies.

Methods

Experimental Setup and Procedures

Schematic diagram of the HDH system used in this study is shown in Fig. 1. The details of the HDH system and acoustic measurement setups can be found in our previous publication [4]. A fresh non-fixed human post mortem temporal bone was used for the measurements. The cartilaginous and boney parts of the ear canal were removed to provide optical access to the TM. The middle-ear cavity was opened to check the middle ear ossicles' healthiness, and then it was sealed using the sealing cement. A small metal tube was installed through the Eustachian tube for injecting fluid into the middle ear cavity. The TM surface was sprayed with a thin layer of HOLBEIN water soluble oil color (TITANIUM WHITE) using a MASTER airbrush (Model S68).

After the control measurement on the normal ear, the same sample was subjected to various manipulations simulating different middle ear pathologies. The shape and transient acoustic induced displacements were measured at each manipulation (experimental setup is shown in Fig. 2a). The manipulations include: open middle ear cavity, close middle ear cavity, inject half TM in contact with fluid (saline) (Fig. 2b), full TM in contact with fluid, remove the injected fluid; fix the footplate of the

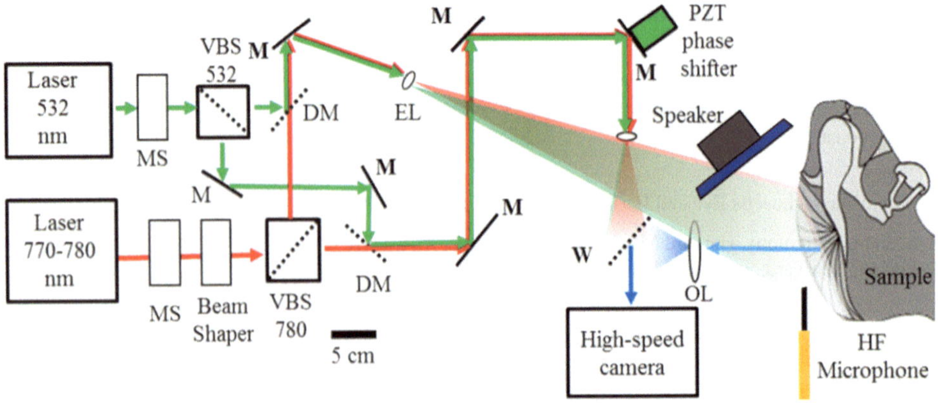

Fig. 1 Schematic of High-speed Digital Holographic (HDH) system [4]. Laser at 532 nm is used for displacement measurements, and 770–780 nm tunable Laser is used for shape measurement. Two lasers are coupled into the same optical path to ensure the shape and displacement measurements are in the same coordinate system minimizing the efforts for image registration

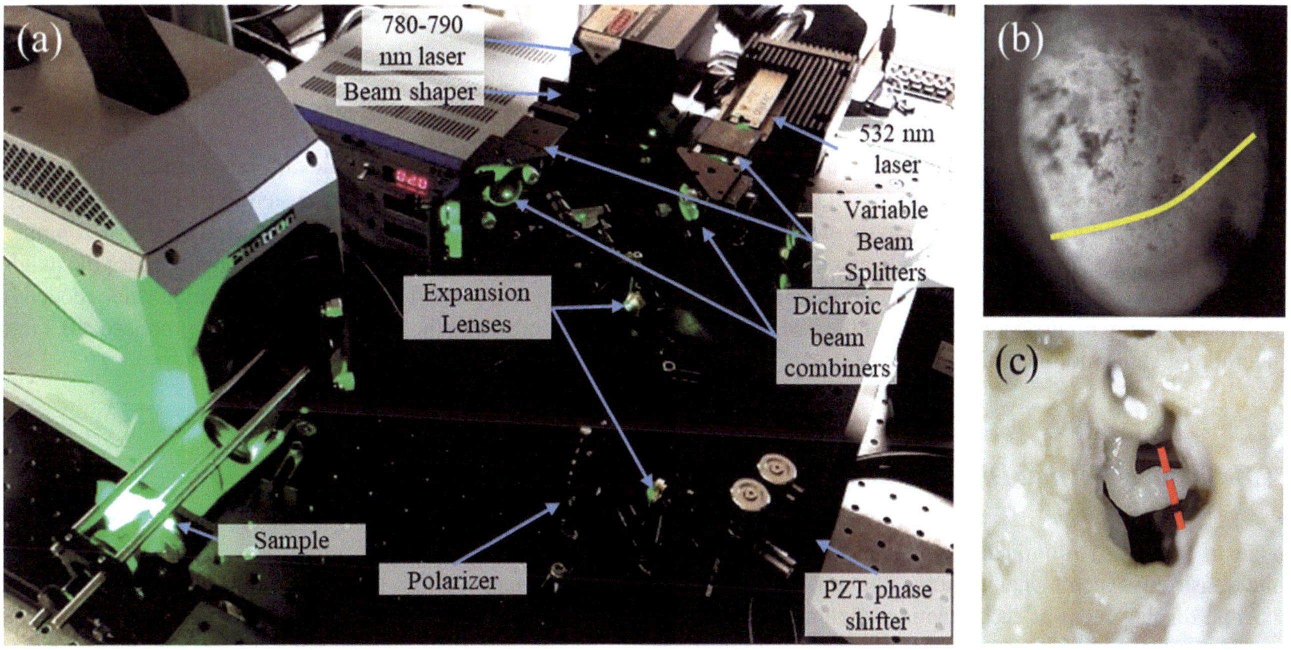

Fig. 2 (**a**) shows the setup, measuring instrumentations and the sample; and (**b**) is a photograph of the human post-mortem TM with half level saline injected. The yellow contour shows the level of fluid. (**c**) shows Incudo-stapedial (IS) joint. The red dash shows where the IS joint is interrupted

stapes by super glue; and interrupt the Incudo-stapedial (IS) joint (Fig. 2c). The transient responses due to a short acoustic click (i.e., a 50 μs square pulse) are compensated by the shape information to calculate the true out-of-plane motion of the entire TM [4].

Experimental Modal Analysis

Frequency Response Function (FRF) matrices are calculated for all the TM surface points by dividing the displacement by the sound pressure stimulus (monitored by a high frequency microphone next to the TM surface, see Fig. 1) in the frequency domain at various middle ear manipulations. Complex mode indicator functions (CMIFs) are calculated by taking singular value decomposition ([5], Eq. (1)) of each FRF matrix,

$$[H(\omega)] = [U(\omega)]\ [\Sigma(\omega)]\ [V(\omega)]^H, \tag{1}$$

where H(ω) is the frequency response function matrix; U(ω) is the left singular vector matrix (unitary); Σ(ω) is the singular value matrix (diagonal); V(ω) is the right singular vector matrix (unitary); $[\ ]^H$ marks the Hermitian (conjugate transpose) of matrix. The diagonal axis of the singular value matrix Σ(ω) are plotted in logarithmic scale to visualize and compare the frequency responses from different tests.

The Impulse Response Functions (IRF) are calculated by converting the FRFs back to the time domain. The least-squares complex exponential method is then applied to the IRFs to find the complex system poles, and then natural frequencies and modal damping are derived from the complex poles [6]. Unphysical poles (poles without a complex conjugate pair or poles with non-negative real components) are rejected. Identification of complex modes depends on the convergence of natural frequency and modal damping estimates while increasing the number of possible poles. Mode shapes are calculated using the least-squares frequency domain method based on the identified structure poles [7]. Modal Assurance Criterion (MAC) matrices extracted from mode shapes, weighted by the natural frequency and damping of each mode, are used to quantitatively compare control and different pathological conditions [8].

Representative Results

TM Shape Comparisons

The 3D shapes of the TM under different middle ear manipulations are compared to see how different manipulations affect the TM shape. Figure 3a shows a typical 3D shape rendering of the TM measured by HDH system, and Fig. 3b shows 2D maps of the TM under the normal (cavity closed) and Incudo-stapedial (IS) joint interrupted conditions (left two columns) and compares the shape difference between the two conditions (right most column). Color bar in each panel codes the height of the TM point relative to the TM annulus (0 mm as a reference). The center of the TM has the largest height value (coded in red), consistent with the conical shape of the TM. Figure 3b clearly demonstrates the IS joint interruption produced a 0.3 mm lateral (outward) motion of the umbo and center of the membrane.

TM Surface Normal Displacement

Due to near field illumination and the observation configuration of the HDH system, there is a significant change in the sensitivity vector change along the surface of the TM and the shape of the TM is used to calculate the surface normal displacement at each location to compensate for the sensitivity vector differences [4]. Once the sensitivity vector is compensated by the shape information, full-field displacements along the TM's surface normal direction in response to acoustic click excitation under normal middle ear condition and displacement time waveforms at six individual locations on the TM surface are shown in Fig. 4. The top panel of Fig. 4b shows the acoustic waveform monitored by the microphone. One can see different regions of the TM have different response time relative to the acoustic signal, which might be explained by the delay due to wave traveling along the TM surface.

Complex Mode Indicator Function (CMIF)

To visualize and compare all the data of the entire surface of the TM across different manipulations, the complex mode indicator function (CMIF) is used as a generalized presentation of the membrane's frequency response (Fig. 5). The horizontal axis is the frequency axis in kilohertz, and the vertical axis is the singular value in decibel which is the square roots of the eigenvalues of the Frequency Response Function (FRF) matrix. We observed that CMIF changes with different middle ear conditions. Figure 5 shows CMIFs computed from one temporal bone with different middle ear manipulations. If the cavity close condition (black) is considered as a baseline, cavity open condition (red) has lower CMIF amplitude at low frequencies and an additional major peak at 3.6 kHz. Half fluid (yellow) and full fluid (dark blue) conditions have significantly lower amplitude across the entire frequency spectrum compared to the cavity closed with the full fluid condition producing the lowest CMIF amplitudes of the data set. Stapes fixation condition (cyan) has lower amplitude in low frequency range (0.5–2 kHz) and a less significant peak at 2.1 kHz, whereas the IS joint interruption condition (green) has a sharp peak at 10 kHz.

Modal Damping and Natural Frequency

The computed modal damping at identified natural frequencies for each middle-ear condition displays a general decay trend as frequency increases. A comparison between control (cavity close) and Stapes fixation and IS joint interruption are shown in Fig. 6. The identified natural frequencies of the middle ear system are different between normal and pathological ears (blue lines with dots do not overlap with red lines with dots). The fitted modal damping curves (blue and red lines) suggest the modal damping is reduced in the lower frequency range (1–4 kHz) in the Stapes fixation case (Fig. 6a). The modal damping of the IS joint interruption case is lower than the normal condition across the entire frequency range (Fig. 6b).

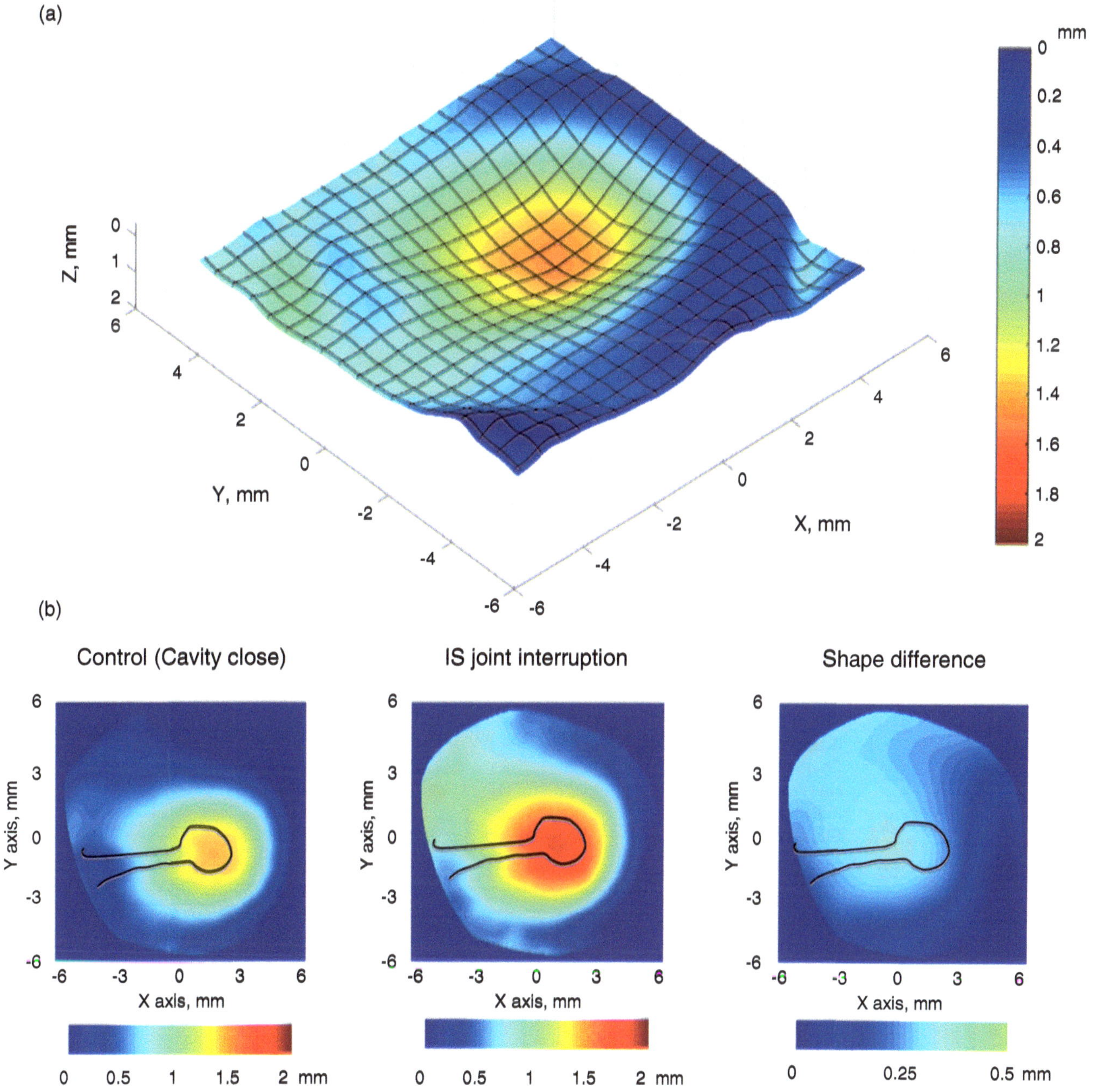

Fig. 3 Representative HDH shape measurement results. (**a**) 3D illustration of TM shape measured by HDH system; (**b**) shape comparison between normal condition (cavity closed) and IS joint interruption condition. The shape comparison shows the HDH system has the resolution to detect the change of the TM's shape due to the manipulation

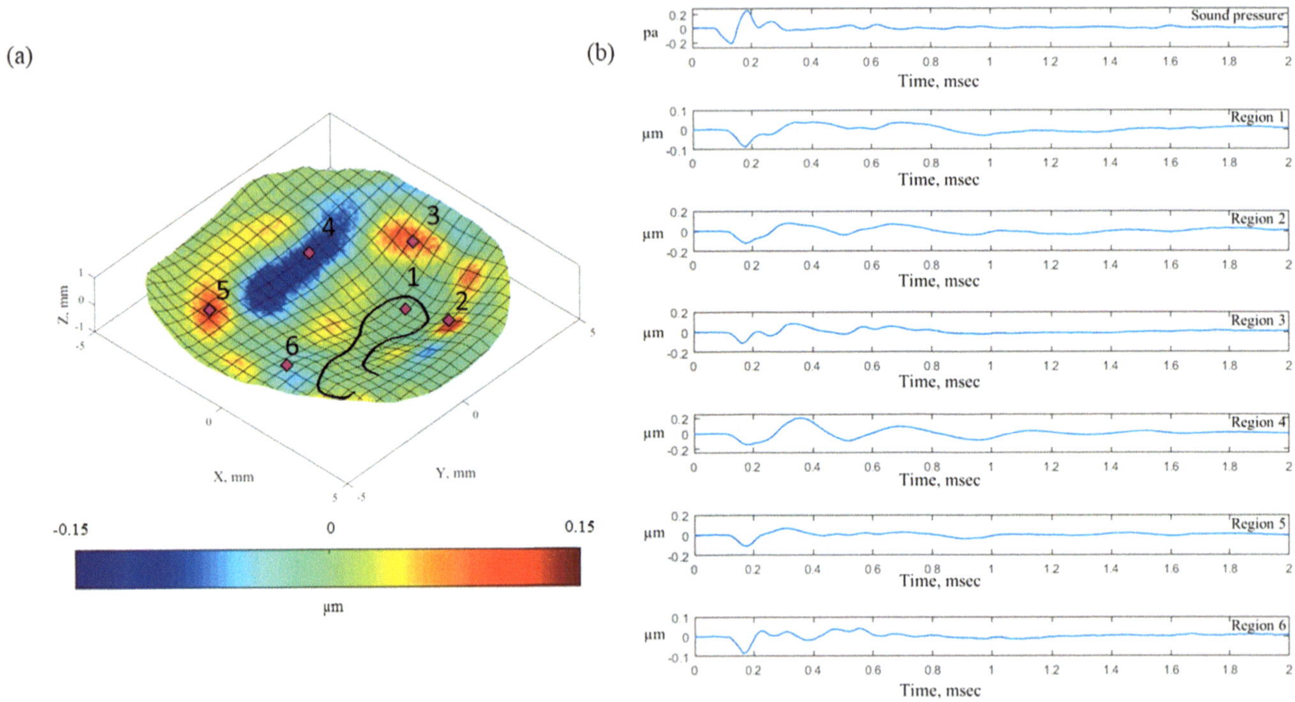

Fig. 4 Time waveforms at six selected regions across the TM for acoustic click measurements. (**a**) Transient displacements after 300 μs of acoustic excitation. Displacement time waveforms of marked points are shown in (**b**); (**b**) acoustic signal measured by high frequency microphone and displacement time waveforms at six selected regions

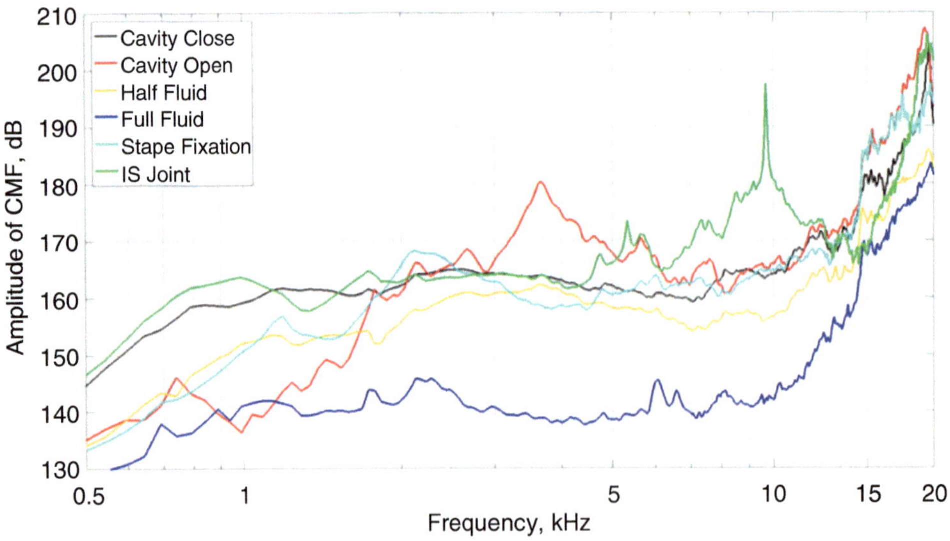

Fig. 5 Complex mode indicator function of the middle ear system under different manipulations. Each complex mode indicator shows some unique characteristics. However, the significance of these differences will depend on how repeatable they are in the different test samples and the physical meaning of the complex mode indicator should be investigated

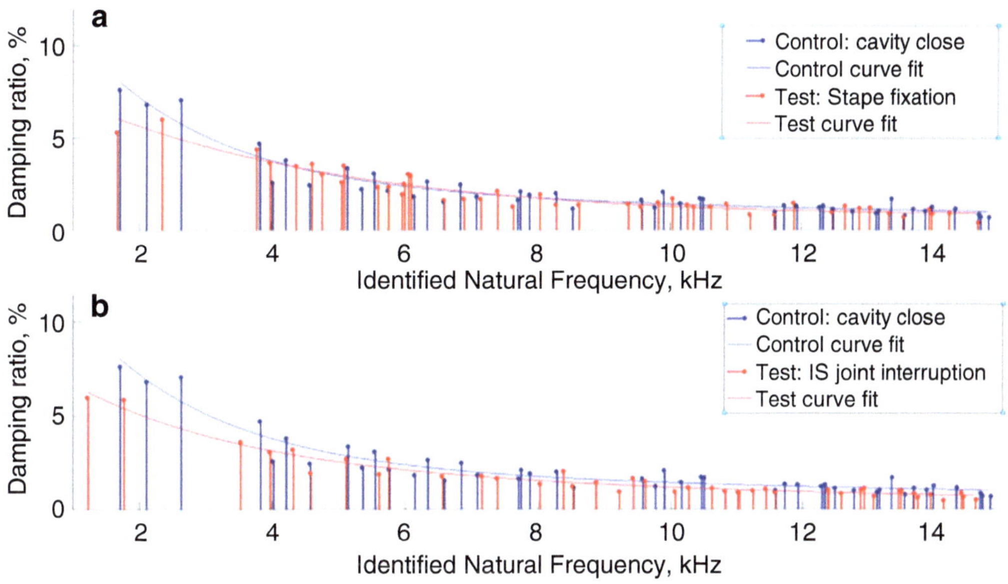

Fig. 6 Modal damping comparison (**a**) compares cavity closed and Stapes fixation; (**b**) compares cavity closed and IS joint interruption. The modal damping ratio decays as frequency increases and the manipulation affects TM's modal damping as well as its natural frequencies

Modal Assurance Criterion (MAC)

Figure 7 shows the MAC correlation coefficients between the control case (cavity closed, X-axis) and various middle ear manipulations (Y-axis). MAC correlation coefficient of value 1 (black color in color bar) shows a perfect similarity between the two modes. However, in practice, values greater than 0.6 (orange color) indicate that the modes are highly correlated. The top two panels in Fig. 7 suggest nearby modes in the cavity open and half fluid conditions are moderately correlated (orange bars at many modal frequencies) with those in the cavity-closed condition. However, rest of panels in Fig. 7 suggest full fluid injection, stapes fixation and IS joint interruption have significant lower correlation coefficients (white or yellow bars), except a few sparse high correlations at the lowest or the highest modal frequencies in Stapes fixation and IS-joint interruption cases.

Conclusions and Future Works

Our results provide by far the most complete description of TM mechanics including shape information, vibration motion and modal parameters such as natural frequency, modal damping, and mode shape. Experimental Modal Analysis (EMA) results can separate different middle ear pathologies, suggesting the potential for clinical diagnosis. In the future works, we will continue to verify the accuracy of EMA results through further experiments; identify trends in the data associated with different pathologies as well as test the sensitivity and selectivity of these analyses for clinical diagnosis. Artificial Intelligence (AI) and Data Mining will be applied to automate the EMA process and assist in the separation of normal and diseased states.

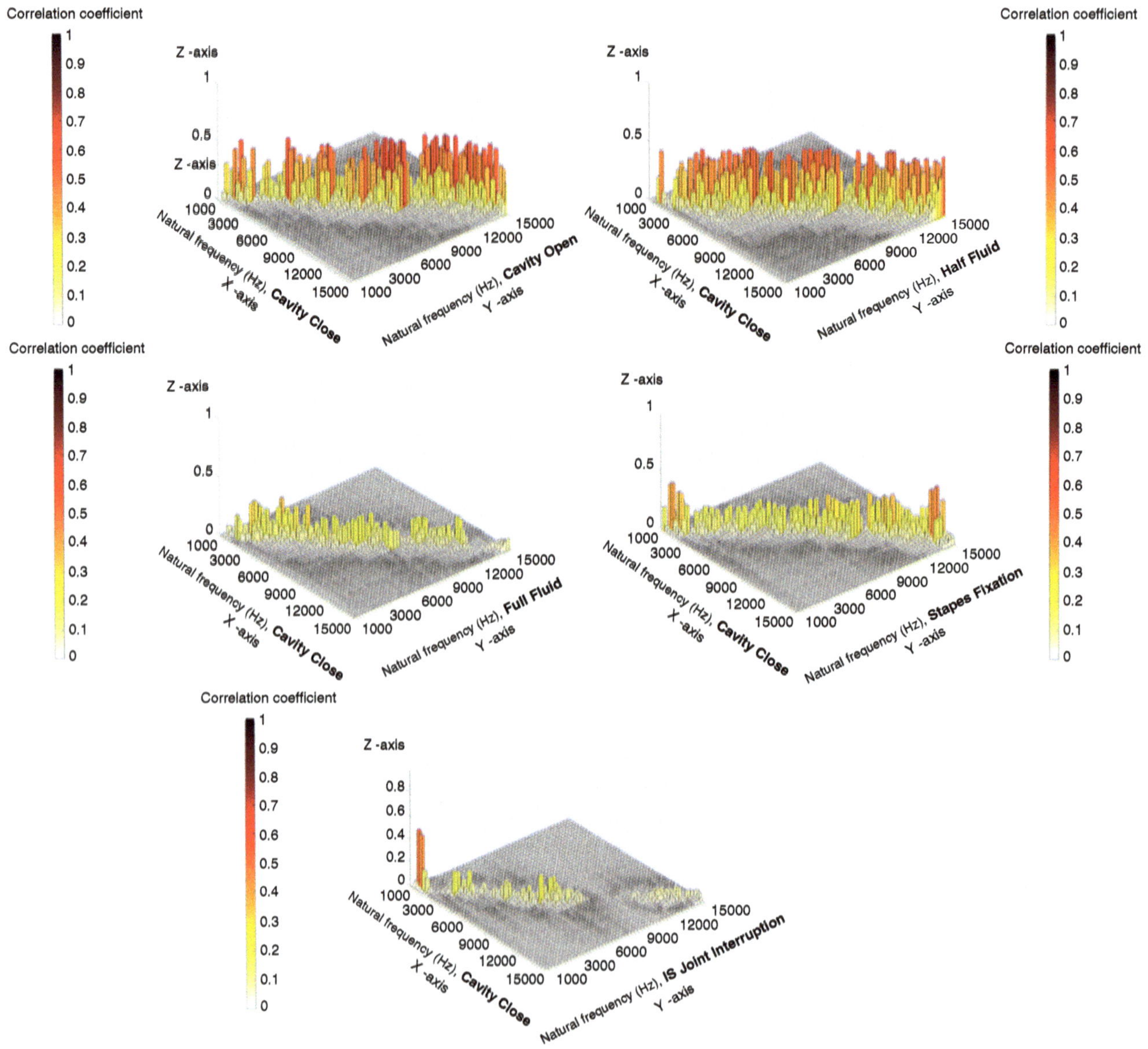

Fig. 7 Mode shape comparison by Modal Assurance Criterion (MAC) between cavity closed and various middle ear pathological conditions. MAC is a good statistic indicator of the consistency between mode shapes. The higher the correlation coefficient indicates similar mode shapes. Cavity open and half fluid are similar to the control (cavity closed) compare to full fluid, stapes fixation and IS joint interruption

References

1. P. Razavi, J.T. Cheng, C. Furlong, J.J. Rosowski, High-Speed Holography for In-Vivo Measurement of Acoustically Induced Motions of Mammalian Tympanic Membrane, in *Mechanics of Biological Systems and Materials, Proceedings of the SEM*, 2016
2. I. Dobrev, C. Furlong, J.T. Cheng, J.J. Rosowski, Full-field transient vibrometry of the human tympanic membrane by local phase correlation and high-speed holography. J. Biomed. Opt. **19**(9), 96001 (2014)
3. M. Khaleghi, C. Furlong, J.T. Cheng, J.J. Rosowski, Thin-Shell Behavior of Mammalian Tympanic Membrane Studied by Digital Holography, in *Mechanics of Biological Systems and Materials*, ed. by F. Barthelat, C. Korach, P. Zavattieri, B. Prorok, K. Grande-Allen, vol. 7, (Springer, Cham, 2015), pp. 19–25
4. P. Razavi, H. Tang, J. Rosowski, C. Furlong, J.T. Cheng, Combined high-speed holographic shape and full-field displacement measurements of tympanic membrane. J. Biomed. Opt. **24**(3), 031008 (2018)
5. R.J. Allemang, D.L. Brown, A Complete Review of the Complex Mode Indicator Function (CMIF) with Applications, in *Proceedings of ISMA International Conference on Noise and Vibration Engineering*, 2006
6. D.L. Brown, R.J. Allemang, R. Zimmerman, M. Mergeay, Parameter estimation techniques for modal analysis, in SAE Technical paper 790221, 1979
7. T.K. Caughey, M.E.J. O'Kelly, Classical Normal modes in damped linear dynamic systems. J. Appl. Mech. **32**, 583–588 (1965)
8. P. Vacher, B. Jacquier, A. Bucharles, Extensions of the MAC criterion to complex modes, in *Proceedings of ISMA2010*, 2010, pp. 2713–2726

Influence of Cell Wall Polysaccharides on Structure and Mechanics of *Streptococcus mutans*

Joree N. Sandin, Natalia Korotkova, and Martha E. Grady

Abstract *Streptococcus mutans* (*S. mutans*) is a group of cocci bacteria that highly contributes to oral decay. In cases of high potency, it can cause plaque build-up, impaired speech, difficulty chewing, and cavity formation, which contribute to over half of the dental visits in the United States. In severe cases where a biofilm develops on a dental implant, patients will experience pain, swelling, and potential loosening or loss of the titanium implant. *S. mutans* are also highly resistant to antibiotics, which makes the infections both persistent and difficult to treat. As poor oral hygiene continues to be a global epidemic, it is important to study and characterize the mechanics of the bacteria to develop therapeutic targets to alleviate *S. mutans* infections. One possible target is modifications of the cell wall. Recent literature suggests that alterations to the biosynthesis pathways of cell wall polysaccharides could lead to new opportunities for therapeutics. The cell wall of *S. mutans* is high-functioning and complex. It consists of multiple peptidoglycan layers, wall teichoic acid (WTA) containing surface glycopolymers, and a polysaccharide capsule. Using atomic force microscopy (AFM) in combination with fluorescent laser scanning confocal imaging, we compare cell wall deformation with mutants defective in WTA. Furthermore, scanning electron microscopy of both wild type *S. mutans* and mutant strains reveal differences in surface morphology. Our long-term goal is to determine mechanical and structural properties of bacterial cell walls that contribute to antibiotic resistance in order to preferentially regulate such properties.

Keywords Atomic force microscopy · Confocal microscopy · *Streptococcus mutans* · Cell wall · Bacteria

A wild type strain of *S. mutans* was selected first to study to understand mechanics of the naturally occurring form. To begin developing therapeutics to treat infection, a mutant type strain was also developed and studied for comparison. Manipulation of the wall teichoic acid layer and modification of the glycerol phosphate affects the properties that the bacteria exhibit, therefore affecting the cell wall mechanics of the mutant strain. *S. mutans* are gram-positive bacteria that have a thick cell wall composed of the polysaccharide, glycopolymer, and peptidoglycan layers [1]. For control and better understanding, a third strain of bacteria was introduced into the experiment, which is a complemented form generated by the deletion of the gacH, a gene with unknown function [1]. Individual morphology [2], surface roughness, and elastic moduli will be characterized and studied across these three strains *S. mutans*.

Employing AFM in combination with confocal microscopy enables improved insight into cell wall mechanics of *S. mutans*. Since these bacteria are only a micron in length, on average, the tools that can be used to study the mechanical and structural properties effectively are extremely limited. Before the introduction of AFM into microbiological applications, traditional light microscopy was used to view and studied cells and bacteria [3]. However, now, using AFM, more detail and information can be collected on cultured samples that includes morphology, cytoskeleton characteristics, migration patterns, locomotion, organelle movement, cell volume, etc. [3]. Research on cell wall mechanics requires the ability to culture the bacterial strains in a sterile environment using appropriately-sized glass-bottomed petri dishes, stain with a dye having proper excitation and emission spectra, observe using high-magnification oil objectives in an inverted manner, laser scan for bacterial thickness in planktonic form, and, finally, probe using AFM to collect force indentation curves to extract cell wall mechanical properties.

J. N. Sandin · M. E. Grady (✉)
Department of Mechanical Engineering, University of Kentucky, Lexington, KY, USA
e-mail: joree.sandin@uky.edu; m.grady@uky.edu

N. Korotkova
Department of Molecular and Cellular Biochemistry, University of Kentucky, Lexington, KY, USA
e-mail: nkorotkova@uky.edu

M. E. Grady (ed.), *Mechanics of Biological Systems and Materials & Micro-and Nanomechanics, Volume 4*,
Conference Proceedings of the Society for Experimental Mechanics Series, https://doi.org/10.1007/978-3-030-30013-5_12

Previous studies of the use of AFM on biological samples have helped motivate our research [4–7]. AFM was invented in 1986 by Binning et al., and has advanced greatly since [5]. It's first application was for creating nanodevices in semiconductors but has only recently become so prevalent in the microbiological area. AFM has been adapted to characterize living cell topography as well as mechanical properties [7–10] with nanoscale precision. AFM can be performed in buffer maintaining biological conditions and allowing simultaneous imaging from below. The first two classical AFM modes are known as contact mode and tapping mode [11, 12]. While using contact mode, the AFM tip will scan over the respective sample by resonating at a given frequency, based on the properties of the manufactured tip, to raster high resolution images in terms of the sample's height. When changes in deflection occurs, a laser, that is pointed at the cantilever reflective tip, bends accordingly and the position-sensitive detector collets the data. For each indentation into the sample, a force curve is generated as a function of deflection to give information on the stiffness and adhesion characteristics of a particular area on the sample. Contact mode maintains a constant vertically applied force of constant magnitude. In tapping mode, the tip oscillates at its respective resonance frequency to scan over the sample, maintaining a constant amplitude damping, at intermittent contact [3]. Due to the high stiffness nature of the tip used for this mode, the change in amplitude of an oscillation during the raster provides the information of the sample surface topography. However, it is important to keep in mind that although being highly effective for non-optical microscopy, care must be taken during AFM on biological samples to ensure tip compatibility [3, 13]. For example, a sharp tip brought into contact with a living cell at a fast approach velocity can puncture the cell. Therefore, topographic information is not desirable when using contact mode, in which case, tapping mode is more typically used [11].

JPK-Bruker atomic force microscopes come with another new mode known as Quantitative Imaging™ mode (QI™). QI™ mode allows for the scanning and data collection over samples that are difficult to measure and that exhibit very soft qualities, while still producing a high-resolution image. Studies on QI™ mode have shown to be useful on samples that are immobilized or loosely immobilized and allows for the collected data to be processed and analyzed using multiple models, which is a strong advantage [11].

Images of single bacteria immobilized on glass surfaces were successfully obtained by AFM. Protocol for performing confocal in conjunction with AFM requires bacteria cell culture, immobilization, staining, and imaging. Three strains of *S. mutans* were grown overnight in Todd Hewitt Yeast (THY) with the addition of antibiotics for all strains. One mutant type and the complemented form strains also had an addition of chloramphenicol (2.5 μg/mL), and all dishes were incubated overnight. The next morning, 30 mL of THY was inoculated with 3 mL of bacterial suspension. When culture reached an optical density of 0.6–0.7, 10 mL of the bacterial culture was spun for 10 min at 3900 rpm on $+4^{\circ}$ to discard the supernatant. Next, the pellet was suspended in 10 mL of PBS and spun for 10 min at 3900 rpm on $+4^{\circ}$ to, again, discard the supernatant. While spinning, World Precision Instruments (WPI) 35 mm dish diameter, 23 mm glass diameter, 0.17 mm glass thickness, FluoroDish™ Tissue Culture with Glass Bottom dishes were covered with poly-l-lysine solution. The dishes were selected from an approved dishes list for the JPK-Bruker Nanowizard® 4a AFM. Then, the dishes were incubated for 1 h at room temperature in a biosafety cabinet. The poly-l-lysine solution was removed, and each dish was rinsed three times with PBS. The glass surfaces were left to dry for 1 h at room temperature in the biosafety cabinet. Then, the pellet was resuspended in 1 mL of 20 mM HEPES (pH of 8.0 + 0.5% BSA). Two centrifuge tubes were used to add 0.5 mL of suspension to each, then 0.45 mg/mL of AtlA protein was added to the first tube and 0.45 mg/mL of GFP was added to the second (control) tube. After substrates finished drying, the suspension was transferred onto the prepared glass dishes, and incubated at 37°C for 1 h. After incubating, each dish was washed three times with PBS, then fixed with 4% paraformaldehyde for 15 min in the incubator. Once the final incubation process was complete, each dish was gently washed three times with PBS, and left with all liquid aspirated out in a refrigerator. Now ready, the dishes were transferred to the Light Microscopy Core for rehydration using about 1 mL of PBS before imaging. We began by loading a sample into the cell heater stage, which keeps the bacteria at 37 °C. Using a 20× Nikon air objective, the z-height for the focal plane of bacteria is noted. The sample is removed, and a cantilever tip is loaded into the AFM head. For QI™ mode, on planktonic S. *mutans*, we use Mikromasch tip HQ:CSC17/Cr-Au BS, with tip radius of 8 nm, force constant of 0.18 N/m, and resonance frequency of 13 kHz. After loading the tip, it is lowered into the focal plane until it is sharp and centered, at which the z-height, length, and width of the cantilever is noted. Then, the JPK laser is turned on and positioned such that it is aimed at the reflective side of the cantilever where the tip has been grown. Next, the cantilever is retracted, and the objective is switched to a Nikon 100× oil. Microscope immersion oil is placed on the objective and vacuum grease is dabbed onto the outer rim of the cell dish heater. The sample is once again placed onto the stage and stabilized by the grease. The objective is raised up to the bottom of the glass FluoroDish™ until the oil wicks. The AFM head is placed back onto the stage and the cantilever is lowered into the sample until it appears sharp in the viewing pane. At this point, we run a non-contact calibration in water using the JPK software, taking note of the spring constant and sensitivity. Once calibration is complete, the sample is ready to be scanned by the AFM. Our procedure is to conduct three types of scans, repeated in three to five different locations on each dish: one scan, 5 μm × 5 μm, containing many bacteria, a second scan, 1.5 μm × 1.5 μm, containing a single bacterium, and a final

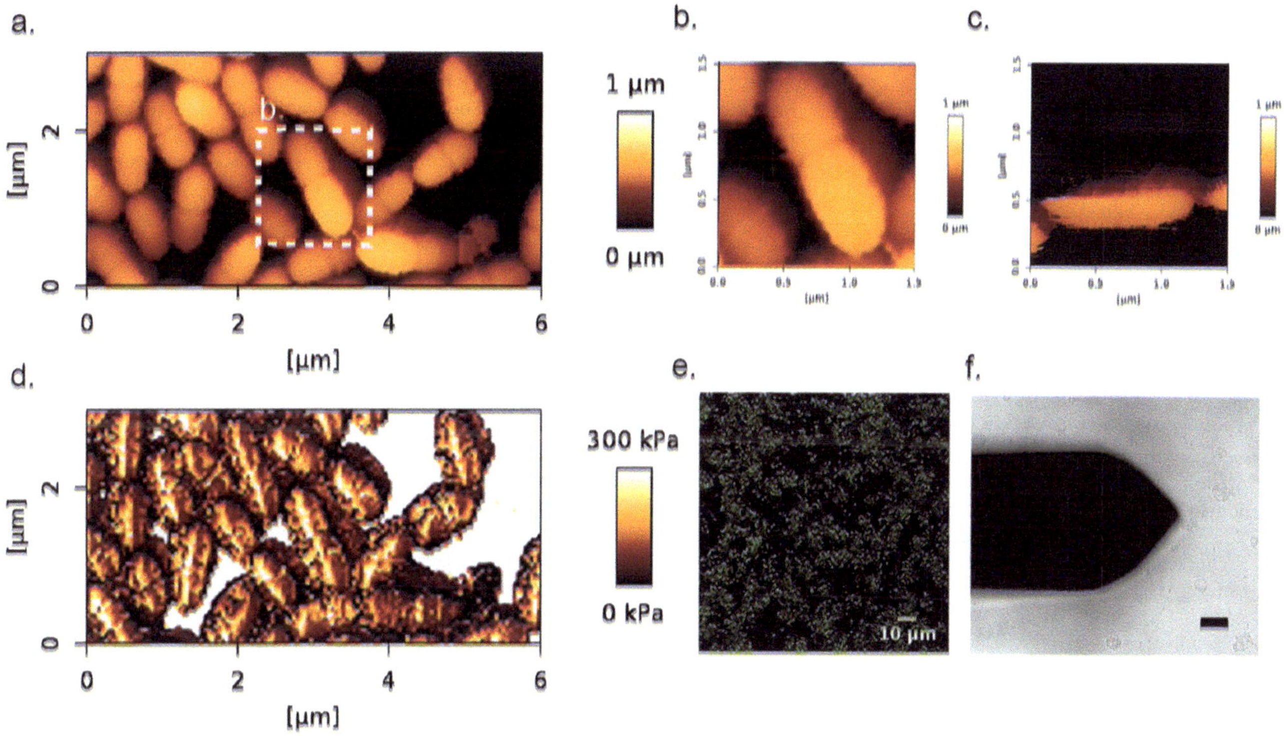

Fig. 1 (**a**) AFM image of *S. mutans* surface profile. (**b**) Highlighted region denoted by dashed line in (**a**) included with more detail on single bacterium. (**c**) AFM image of *S. mutans* indicating poor immobilization procedure. (**d**) AFM modulus map of *S. mutans*. (**e**) Scanning confocal image of *S. mutans* (green). (**f**) AFM cantilever from Mikromasch used with QI™ mode. Scale bar is 10 μm

scan, 300 nm × nm, on the surface of a single bacterium. From this procedure, we collect force curves in which we can extrapolate moduli maps, height measurement maps, and surface roughness values.

Optimized protocols for bacteria cell culture, performing confocal microscopy, and performing AFM are essential for proper aseptic technique, consistency, and effective sample making and testing, which contributes to the overall data collection. Figure 1a is an AFM scan of the wild type *S. mutans* strain, which shows a surface profile of the cell wall and success of AFM on bacteria cells. Here, septal regions where bacteria binary fission is occurring is very prevalent. When culturing bacteria, the protocol for immobilization is paramount [14]. In Fig. 1b, c we compare two single cell bacteria to each other. Figure 1b, is a bacterium from Fig. 1a showing a surface profile where immobilization was successful. In Fig. 1c a bacterium, also showing a surface profile, where immobilization was unsuccessful is seen. When samples are poorly immobilized, the cells tend to move freely, either in the liquid or when the AFM tip begins to contact and scan over the sample, resulting in AFM images that have incomplete cell scans and poor data quality. However, when immobilization is successful, AFM measurements of single bacteria can be achieved and well defined, such as in Fig. 1b. Elastic moduli maps are also generated by the JPK Data Processing software (seen in Fig. 1d). These figures show the respective moduli values on top of *S. mutans* bacteria, giving insight into the soft properties of the sample. Laser scanning confocal imaging was successful, as seen in Fig. 1e which provides more information on the growth pattern and fluorescent properties of *S. mutans*. Finally, an image of the Mikromasch cantilever used in QI™ mode is shown in Fig. 1f. In this brightfield image, taken on the Nikon A1 microscope, small clusters of *S. mutans* are seen in the back focal plane, showing the vast difference between the size of bacteria and the tip we use to scan for mechanical properties, once again, proving the small-scale nature of this research. Using this data, the cell wall characteristics of the four types of strains can begin to be better understood. Knowing the mechanical properties and applying the biological properties of the individual strains will aid in explaining and understanding why the wild type is highly antibiotic resistant to toxic compounds, its morphological properties, how it generates destructive biofilms, and their role in neutrophil killing. These qualities are pertinent for the future development of antibiotics and therapeutics to prevent the associated illnesses that are developed from *S. mutans* infection.

Acknowledgements We thank the Department of Molecular and Cellular Biochemisty for use of bacterial culture equipment and materials. AFM and confocal microscopy were carried out in the UK Light Microscopy Core user facility. The facility provides access to and expertise in a wide range of advanced imaging instrumentation that include confocal, multiphoton, and super resolution microscopes. This work was also supported by an Igniting Research Collaboration Program at the University of Kentucky.

References

1. R.J. Edgar, V.P. van Hensbergen, A. Ruda, A.G. Turner, P. Deng, Y. Le Breton, N.M. El-Sayed, A.T. Belew, K.S. McIver, A.G. McEwan, A.J. Morris, G. Lambeau, M.J. Walker, J.S. Rush, K.V. Korotkov, G. Widmalm, N.M. van Sorge, N. Korotkova, *Discovery of glycerol phosphate modification on streptococcal rhamnose polysaccharides*. bioRxiv 337519 (2019)
2. R.D. Turner, S. Mesnage, J.K. Hobbs, S.J. Foster, Molecular imaging of glycan chains couples cell-wall polysaccharide architecture to bacterial cell morphology. Nat. Commun. **9**(1), 1263 (2018)
3. S.S. Schaus, E.R. Henderson, Cell viability and probe-cell membrane interactions of XR1 glial cells imaged by atomic force microscopy. Biophys. J. **73**(3), 1205–1214 (1997)
4. C. Formosa-Dague, R.E. Duval, E. Dague, et al., Semin. Cell Dev. Biol. **73**, 165–176 (2018)
5. S. Liu, Y. Wang, Application of AFM in microbiology: A review. Scanning **32**(2), 61–73 (2010)
6. V.D. Gordon, M. Davis-Fields, K. Kovach, C.A. Rodesney, Biofilms and mechanics: A review of experimental techniques and findings. J. Phys. D. Appl. Phys. **50**(22), 223002 (2017)
7. S.A. James, L.C. Powell, C.J. Wright, Atomic force microscopy of biofilms—Imaging, interactions, and mechanics. in *Microbial Biofilms-Importance and Applications* (InTech, 2016)
8. A. Cartagena, A. Raman, Local viscoelastic properties of live cells investigated using dynamic and quasi-static atomic force microscopy methods. Biophys. J. **106**(5), 1033–1043 (2014)
9. A. Beaussart, C. Pechoux, P. Trieu-Cuot, P. Hols, M.-Y. Mistou, Y.F. Dufrene, Molecular mapping of the cell wall polysaccharides of the human pathogen Streptococcus agalactiae. Nanoscale **6**(24), 14820–14827 (2014)
10. Y.J. Oh, B. Plochberger, M. Rechberger, P. Hinterdorfer, Characterizing the effect of polymyxin B antibiotics to lipopolysaccharide on Escherichia coli surface using atomic force microscopy. J. Mol. Recognit. **30**(6), e2605 (2017)
11. L. Chopinet, C. Formosa, M. Rols, R. Duval, E. Dague, Imaging living cells surface and quantifying its properties at high resolution using AFM in QI™ mode. Micron **48**, 26–33 (2013)
12. G. Smolyakov, C. Formosa-Dague, C. Severac, R. Duval, E. Dague, High speed indentation measures by FV, QI and QNM introduce a new understanding of bionanomechanical experiments. Micron **85**, 8–14 (2016)
13. E.P. Wojcikiewicz, X. Zhang, V.T. Moy, Force and compliance measurements on living cells using atomic force microscopy (AFM). Biol. Proced. Online **6**(1), 1–9 (2004)
14. L. Kailas, E. Ratcliffe, E. Hayhurst, M. Walker, S. Foster, J. Hobbs, Immobilizing live bacteria for AFM imaging of cellular processes. Ultramicroscopy **109**(7), 775–780 (2009)

Dental Implant Texture Affects Biofilm Adhesion Strength

James D. Boyd, Natalia Korotkova, and Martha E. Grady

Abstract Peri-implantitis, a disease formed by subgingival biofilm between dental implants and surrounding tissue, can lead to necrosis or implant loss. The development of an implant surface that promotes osseointegration and deters bacterial biofilm adhesion is paramount to prevent peri-implantitis. A technique to quantify adhesion strengths of biofilms is important to optimize surfaces which prevent bacteria from adhering strongly. The laser spallation technique has been recently adapted to obtain quantitative measures of biofilm adhesion. One key advantage of laser spallation is it results in quantified adhesion strength while using a non-contact high strain rate force. Image analysis can be used to obtain fluence, energy per unit area, at spallation of the biofilm, along with one dimensional wave analysis and finite element analysis, a quantitative interface adhesion strength can be determined for the biofilm-implant interface. In this study, *Streptococcus mutans*, a gram-positive facultative anaerobe, was chosen because it promotes the attachment and growth of more harmful bacteria. The competition between oral bacteria and cell should also be considered when comparing implant surface characteristics. MG-63 was chosen as it closely mimics osteoblast adhesion. We will demonstrate the competition in adhesion between *S. mutans* and osteoblast like cells on dental implant mimicking surfaces through the adaptation of the laser spallation technique. This study will lead to the development of dental implant surfaces which promote osseointegration and inhibit biofilm formation. Furthermore, the laser spallation technique will be used to optimize other medical implant surfaces, and surfaces where biofilms have deleterious effects.

Keywords Biofilms · Laser spallation · Adhesion · *Streptococcus mutans* · MG-63 · Dental implants

According to the American Academy of Implant Dentistry, there are three million people in the U.S. who have implants with an additional 500,000 implanted each year [1]. The attachment of oral bacteria to dental implants leads to peri-implantitis, a disease formed by subgingival biofilm between dental implants and surrounding tissue. Peri-implantitis can lead to the decay of the gum surrounding the implant as well as total failure of the implant [2]. The colonization of several different bacteria associated with peri-implantitis results in an antibiotic resistant organism [3]. Prevention of initial strong biofilm adhesion could alleviate some of the deleterious effects of biofilm-associated implant infections. In order to develop a preventative strategy for oral biofilms, an understanding of factors that contribute to strong biofilm adhesion at implant interfaces must be determined.

Thin film adhesion studies have been conducted in the past through a multitude of techniques including scratch, peel, and stud-pull [4, 5]. During testing with these techniques the interface is subject to very high stress levels with inhomogeneous deformations [6]. As a result, adhesion measurements from these tests are non-quantitative which prevents direct comparison [7]. Laser spallation acts as an adhesion strength test by delivering a photoacoustic shockwave to the biofilm with intent of ablating the film from its applied surface. Spallation testing is followed by a calibration procedure where the amount of stress created at the point of film detachment is measured. This method is superior to other film adhesion testing methods as it is a non-contact method which targets only a specific portion of the sample, leaving the rest of the area unscathed and available for further experimentation. The laser spallation setup used is illustrated in Fig. 1a. Laser spallation has been used in the past to quantitatively measure varying different thin film adhesion strengths [8–11] including mammalian cells [12–15], and more

J. D. Boyd · M. E. Grady (✉)
Department of Mechanical Engineering, University of Kentucky, Lexington, KY, USA
e-mail: james.boyd@uky.edu; m.grady@uky.edu

N. Korotkova
Department of Molecular and Cellular Biochemistry, University of Kentucky, Lexington, KY, USA
e-mail: nkorotkova@uky.edu

M. E. Grady (ed.), *Mechanics of Biological Systems and Materials & Micro-and Nanomechanics, Volume 4*,
Conference Proceedings of the Society for Experimental Mechanics Series, https://doi.org/10.1007/978-3-030-30013-5_13

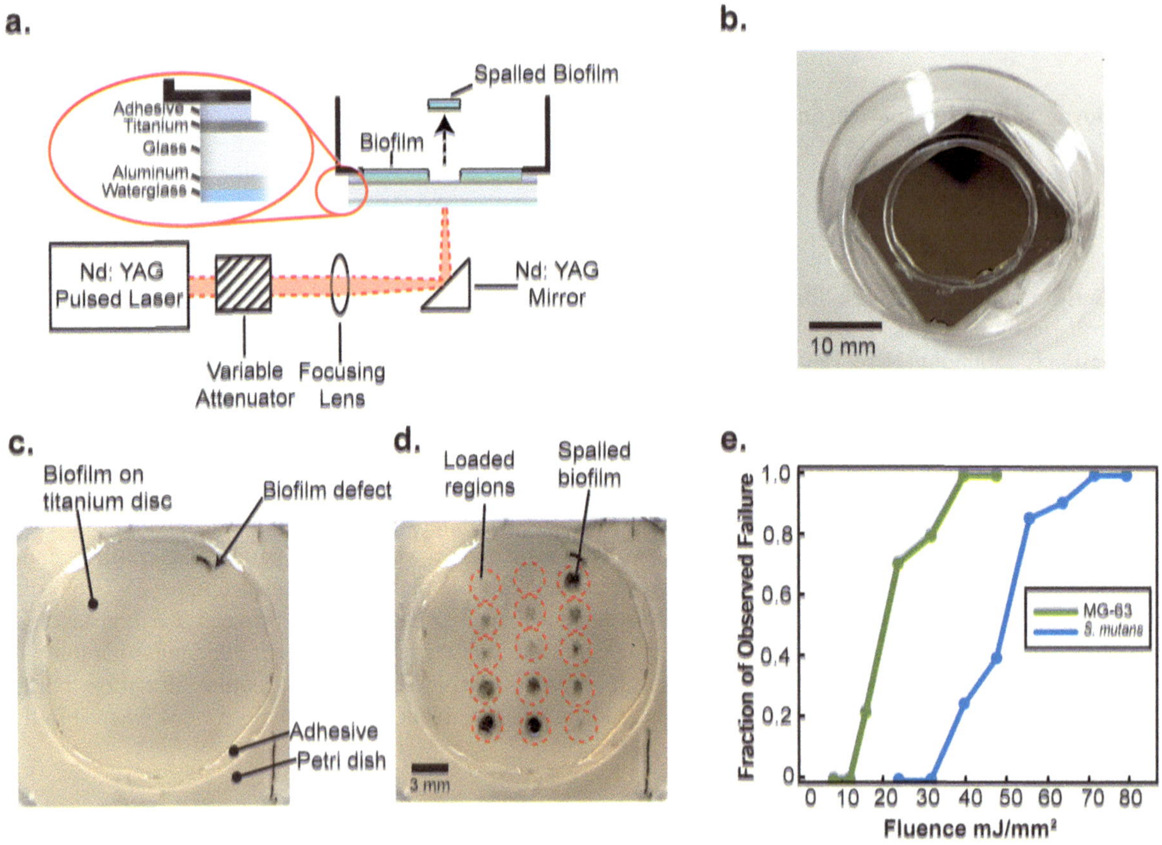

Fig. 1 (**a**) The laser spallation set up used during delamination experiments taken from Boyd et al. The Nd: YAG laser used reflected 90° to allow for horizontal orientation of the substrate assembly used. (**b**) Optical image of typical substrate assembly, which includes the titanium surface before bacterial biofilm colonization. Opposite to the titanium surface is the aluminum energy absorbing layer, not shown in image. *S. mutans* biofilm grown on the titanium surface is shown in (**c**) while the same sample that has been loaded is shown in (**d**). The spallation data was collected for both the MG-63 and the *S. mutans* cultured with 75 mM sucrose and fraction of observed failure are compared (**e**)

recently with bacterial biofilms [16]. Figure 1a depicts the set up used for laser spallation technique with bacterial biofilms and mammalian cells.

In order to mimic peri-implantitis, relevant bacteria must be selected with emphasis on early colonizing bacteria. Early colonizing bacteria are key initiators which promote the adhesion and growth for other harmful microbes onto the fresh implant surface [17]. The initial bacterial colonization process is dominated by oral *Streptococci*, which make up over 80% of early biofilm constituents [18]. *Streptococcus mutans* was specifically selected as a representative early colonizing bacterium. *S. mutans* is known to be a main constituent of dental plaque, and is commonly linked to dental cavities [19].

Successful osseointegration of the dental implant is a pivotal factor for dental implant longevity. *In vivo* osseointegration involves a complex and coordinated action of various cell types. The arrival of osteoblast cells onto the dental implant surface serves as a physiologic barrier to prevent bacteria and other deleterious substances in the oral cavity [20]. In order to mimic osteoblast behavior *in vitro* bone remodeling models have been developed [21]. For our experiments, MG-63 human osteosarcoma cells were used as our bone remodeling model [22]. Spallation experiments have been successfully performed with MG-63 osteoblastic cells and *S. mutans*. The quantitative adhesion measurements collected will enhance the understanding of the cell-bacteria competition on dental implant surfaces. This understanding can be used to fuel smarter implant surface design which prevent bacterial adhesion while promoting cell adhesion and overall successful implant integration.

Test substrate materials must mimic currently used materials within the implant industry. The industry standard dental implant is a commercially pure, 99.995%, titanium alloy, commonly referred to as Ti grade IV. Titanium is commonly used for its mechanical properties, resistance to corrosion, and biocompatibility, along with a prolific use as a superior biomedical implant material for many years [23]. Thus, commercially pure titanium surfaces were used in custom substrate assemblies. The compound substrates include a 1″ × 1″ glass slide with reflective coatings on both sides (100 nm Ti and 300 nm Al). The 300 nm Al layer acts as the energy absorbing layer, this layer converts the laser wave into a mechanical wave to initiate spallation. In order to obtain powerful enough stress waves to initiate delamination, a waterglass layer was added to the backside. The samples were then adhered to the bottom of a 35 mm diameter petri dish using a biologically inert silicone. A 13/16″ hole was cut into the bottom of the petri dish in order to expose the titanium to bacteria and growth media. A typical substrate assembly is imaged in Fig. 1b. A second petri dish was then used to protect the waterglass layer during incubation. The bacteria or cells were then placed inside the petri dish where they are cultured until confluency. This design allows for facile modification of surface characteristics in order to compare the effects on adhesion. Controlling these variables is important and in doing so the goal of developing true interfacial stress of a given biofilm and substrate can be quantified.

Samples were then sterilized, using alchohol and UV exposure, and inoculated with bacteria, with an optical density of 0.7, using Todd Hewitt Yeast broth with 75 mM Sucrose concentration. The sucrose was added to the THY media to increase adhesion strength of *S. mutans*. After successful growth, biofilms were rinsed with phostphate buffered solution (PBS) in order to remove non-adhered bacteria. Substrate assemblies were then mounted horizontally into the laser spallation setup in order to ensure there were no gravitational shearing effects on the biofilm. MG-63 was cultured using Eagle's Minimum Essential Medium, with a seeding density of 120 K cells, over a period of 2 days in order to reach 100% confluecy, with media replaced each day. Similarly cell layers were rinsed with PBS in order to remove non-adhered cells.

Multiple substrate assemblies were tested for biofilm (8 dishes) and cell (6 dishes) films, and multiple locations (12–18) on each film were loaded. The onset of failure of the biofilm-titanium and cell-titanium interface was marked by spallation of the biofilm from the surface of the titanium resulting in a dark spallation region. For example, Fig. 1c, d shows optical images of before and after loading, respectively, for the biofilm at a range of laser fluences.

The graph shown in Fig. 1e depicts the failure statistics determined for both the bacterial and cell adhesion and indicates the overall trend of fluence to rate of spallation for both bacteria and cells. The trend shows that the bacterial-titanium interface requires a greater fluence in order to initiate spallation. In one-dimensional wave mechanics, the stress wave amplitude is directly proportional to film thickness. Assuming biofilm and cells have the same dilatational wave speeds, but different thickness, the thicker film would amplify the stress wave magnitude. Our biofilms are on the order of 10–20 μm thick, whereas cell monolayers are approximately 5–10 μm thick. Therefore, our preliminary data shows a higher fluence required for biofilm removal, which also indicates a higher stress value required for biofilm removal than cell monolayer removal. Additional calibration testing must be conducted in order to convert these fluence values to quantitative interface stresses for the biofilm and cells. Calibration experiments will result in quantitative adhesion strengths, which are directly comparable, for both the bacteria and cell adhesion. Further understanding of the mechanical properties of the *S. mutans* and MG-63 must be determined, confirm film thicknesses and dilatational wave speed, to apply quantitative 1D interface analysis. AFM force spectroscopy measurements are planned to provide these quantitative measures as input into finite element analysis. Future work will investigate the effect of surface roughness on cell and bacteria adhesion. Typical dental implants consist of a sandblasted surface to improve osseointegration. Substrates built with this desired roughness will be tested in order to understand how it impacts the cell-bacteria competition. Here, we have demonstrated the development of the laser spallation technique to compare the adhesion strength of biofilms and cell layers on substrates. With the quantitative information obtained, an accurate comparison can be made between surface characterizations. This initial adhesion baseline of *S. mutans* and MG-63 will show the positive or negative effect of future modifications to the substrate surface. By determining the surface characteristics that promote the adhesion of MG-63 and deter the adhesion of *S. mutans*, we will enable rational selection of dental implant materials for reduced biofilm adhesion and improved clinical outcomes.

Acknowledgements We would like to acknowledge NIH COBRE Phase III pilot funding under number 5P30GM110788-04 to carry out these experiments. We thank the Center for Pharmaceutical Research and Innovation (CPRI) for use of bacterial culture equipment. CPRI is supported, in part, by the University of Kentucky College of Pharmacy and Center for Clinical and Translational Science (UL1TR001998). We thank Dr. Larissa Ponomareva for sharing her bacterial culture expertise. We would also like to thank Drs. Craig Miller, Lina Sharab, and Ahmad Kutkut from the University of Kentucky College of Dentistry for their guidance.

References

1. L. Gaviria, J.P. Salcido, T. Guda, J.L. Ong, Current trends in dental implants. J. Korean Assoc. Oral Maxillofac. Surg. **40**(2), 50–60 (2014)
2. H.J. Busscher, M. Rinastiti, W. Siswomihardjo, H.C. van der Mei, Biofilm formation on dental restorative and implant materials. J. Dent. Res. **89**(7), 657–665 (2010).
3. R. Saini, S. Saini, S. Sharma, Biofilm: A dental microbial infection. J. Nat. Sci. Biol. Med. **2**(1), 71–75 (2011)
4. P.A. Steinmann, H.E. Hintermann, A review of the mechanical tests for assessment of thin-film adhesion. J. Vac. Sci. Technol. A **7**(3), 2267–2272 (1989)
5. K.L. Mittal, Adhesion measurement of thin films. Electrocompon. Sci. Technol. **3**(1), 21–42 (1976)
6. M.D. Thouless, Fracture mechanics for thin-film adhesion. IBM J. Res. Dev. **38**(4), 367–377 (1994)

7. J. Wang, N.R. Sottos, R.L. Weaver, Mixed-mode failure of thin films using laser-generated shear waves. Exp. Mech. **43**(3), 323–330 (2003)
8. M.E. Grady, P.H. Geubelle, P.V. Braun, N.R. Sottos, Molecular tailoring of interfacial failure. Langmuir **30**(37), 11096–11102 (2014)
9. M.E. Grady, P.H. Geubelle, N.R. Sottos, Interfacial adhesion of photodefinable polyimide films on passivated silicon. Thin Solid Films **552**, 116–123 (2014)
10. V. Gupta, A.S. Argon, J.A. Cornie, D.M. Parks, Measurement of interface strength by laser-pulse-induced spallation. Mater. Sci. Eng. A **126**(1), 105–117 (1990)
11. J.L. Wang, R.L. Weaver, N.R. Sottos, A parametric study of laser induced thin film spallation. Exp. Mech. **42**(1), 74–83 (2002)
12. L.L. Hu, X. Zhang, P. Miller, M. Ozkan, C. Ozkan, J.L. Wang, Cell adhesion measurement by laser-induced stress waves. J. Appl. Phys. **100**(8), 084701 (2006)
13. N.C. Francis, I. Kassam, B. Nowroozi, W.S. Grundfest, Z.D. Taylor, Analysis of flexible substrates for clinical translation of laser-generated shockwave therapy. Biomed. Opt. Express **6**(3), 827–837 (2015)
14. P. Miller, L. Hu, J. Wang, Finite element simulation of cell-substrate decohesion by laser-induced stress waves. J. Mech. Behav. Biomed. Mater. **3**(3), 268–277 (2010)
15. H. Nakamura, J. Shim, F. Butz, H. Aita, V. Gupta, T. Ogawa, Glycosaminoglycan degradation reduces mineralized tissue-titanium interfacial strength. J. Biomed. Mater. Res. A **77**(3), 478–486 (2006)
16. J.D. Boyd, N. Korotkova, M.E. Grady, Adhesion of biofilms on titanium measured by laser-induced spallation. Exp. Mech. 1–10 (2018)
17. A.M. Krachler, K. Orth, Targeting the bacteria-host interface: Strategies in anti-adhesion therapy. Virulence **4**(4), 284–294 (2013)
18. J. Kreth, J. Merritt, F. Qi, Bacterial and host interactions of oral Streptococci. DNA Cell Biol. **28**(8), 397–403 (2009)
19. W. Krzyściak, A. Jurczak, D. Kościelniak, B. Bystrowska, A. Skalniak, The virulence of Streptococcus mutans and the ability to form biofilms. Eur. J. Clin. Microbiol. Infect. Dis. **33**(4), 499–515 (2014)
20. R.S. Jayesh, V. Dhinakarsamy, Osseointegration. J. Pharm. Bioallied Sci. **7**(Suppl 1), S226–S229 (2015)
21. E. Czekanska, M.J. Stoddart, R.G. Richards, J.S. Hayes, et al., Eur. Cell Mater. **24**(4), 1–17 (2012)
22. Y.-Y. Chang, H.-L. Huang, Y.-C. Chen, J.-T. Hsu, T.-M. Shieh, M.-T. Tsai, Biological characteristics of the MG-63 human osteosarcoma cells on composite tantalum carbide/amorphous carbon films. PLoS One **9**(4), e95590 (2014)
23. M. Niinomi, Mechanical properties of biomedical titanium alloys. Mater. Sci. Eng. A **243**(1), 231–236 (1998)

On the Role of Human Umbilical Cord Biomechanics

Roberto Brunelli, Massimiliano Papi, Tiziana Parasassi, Marco De Spirito, Carmine Pappalettere, and Luciano Lamberti

Abstract The most abundant component of the human umbilical cord, the Wharton Jelly (WJ), is a mucous connective tissue consisting in a spongy network of interlacing collagen fibers that forms a continuous soft skeleton encasing umbilical vessels. The link between WJ structure and the pivotal cord function in providing unimpeded blood flow to the developing fetus is still poorly understood. We performed an advanced biomechanical characterization of human umbilical cord samples by planar equibiaxial tension tests. The intrinsic moiré technique was utilized for that purpose. A 3-D finite element analysis was then implemented to simulate the biomechanical response of the cord.

Keywords Human umbilical cord · Intrinsic moiré · Mechanical characterization · Anisotropy · Finite element analysis

Introduction

The human umbilical cord is a complex structure that connects the fetus to the placenta and encases the umbilical vessels. The response of its tissues to mechanical loading due to fetal movements and uterine contractions is not well understood. Different studies are reported in the literature on the mechanical characterization of the umbilical cord. Pennati [1] subjected samples of umbilical vein and Wharton's Jelly to uniaxial tensile and stress-relaxation tests. Li et al. [2] determined the biomechanical properties of different segments of human umbilical cord vein. Hamedani et al. [3] compared the mechanical properties of saphenous and umbilical veins. The interest in the mechanical characterization of the human umbilical cord derives from the fact that a strict correlation exists between the health of the neonate and the mechanical behavior of the cord. An umbilical cord not sufficiently stiff does not allow to provide the blood to the fetus at the correct values of pressure and flow. In this study, the intrinsic moiré technique is utilized to measure the in-plane displacement field of 2 mm thick slices of umbilical cord subject to equibiaxial loading. The properties of isotropy/anisotropy of the tested WJ specimens are then assessed. Mechanical response of umbilical cord is then simulated by a detailed 3D finite element model.

Experimental Tests

Ten 2 mm thick slices of fresh umbilical cord were cut in the transverse directions of the cord and submitted to equibiaxial tension. An ad hoc metallic frame was designed and realized to perform the experimental tests (see Fig. 1).

Twelve nylon wires were fixed on the samples through fishing hooks. On the other extremity, three calibrated weights (20 g, 40 g and 60 g, for each wire) were suspended in succession. Pulleys with a 10 mm diameter were utilized to change

R. Brunelli
Dipartimento Materno Infantile e Scienze Urologiche, Università Sapienza, Rome, Italy
e-mail: roberto.brunelli@uniroma1.it

M. Papi · M. De Spirito
Istituto di Fisica, Università Cattolica del Sacro Cuore, Rome, Italy
e-mail: massimiliano.papi@unicatt.it; marco.despirito@unicatt.it

T. Parasassi
Istituto di Farmacologia Traslazionale, Consiglio Nazionale delle Ricerche, Rome, Italy

C. Pappalettere · L. Lamberti (✉)
Dipartimento di Meccanica, Matematica e Management, Politecnico di Bari, Bari, Italy
e-mail: carmine.pappalettere@poliba.it; luciano.lamberti@poliba.it

M. E. Grady (ed.), *Mechanics of Biological Systems and Materials & Micro-and Nanomechanics, Volume 4*,
Conference Proceedings of the Society for Experimental Mechanics Series, https://doi.org/10.1007/978-3-030-30013-5_14

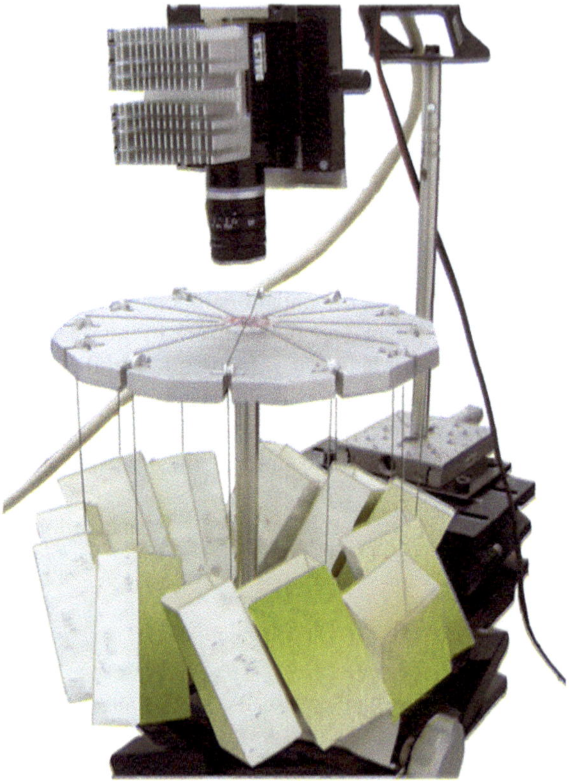

Fig. 1 Experimental apparatus including the loading frame utilized to perform the equibiaxial tests on the umbilical cord sections

the direction of application of the load. By using a timbre, a grating of points with a 500 μm pitch was printed on the samples while a Dalsa CCD camera (1600 × 1360 pixels, 2/3″), equipped with a Fujinon HF9HA-1B objective was utilized to monitor the deformation of the samples. High precision stages were utilized to align the optical axis of the CCD camera with the center of the samples as well as to focus the image of the sample acquired by the camera.

The intrinsic moiré technique [4] was utilized to monitor the modulation of the grating as the specimen deforms under the applied loads. The sensitivity of the optical set-up is the pitch of the grating printed on the specimen.

For each of the acquired images, the two-dimensional FFT was computed. The frequency at which the maximum of the energy in the power spectrum is associated, was identified. Band-pass filters were utilized to select the significant harmonics of the signal; inverse FFT was hence applied to return to the physical space. The phase of the pattern was calculated via in-quadrature filtering. For each point of the sample, the difference between the phase computed in the absence of the load and the phase computed with the load applied is determined. By performing the phase unwrapping operation, it is possible to determine the actual values of in-plane displacements experienced by the sample.

Experimental Results and Material Properties Fitting

Figure 2 shows the grating printed on a sample for different levels of applied load ranging from 0 to 60 g. It is interesting to observe how the application of the load produces a modulation of the dots of the grating printed on the sample surface. Based on this modulation, we can determine the in-plane displacement field of the slice along X and Y-directions.

Figure 3 shows the phase map computed in the reference configuration (i.e. absence of load) and that computed (for the horizontal displacement u_x) for a load of 20 g.

Figure 4 shows the displacement maps measured along directions X and Y, for a load of 20 g applied to all wires. The difference between the spatial domain in which the displacement u_x and u_y was computed derives from the fact that, in order to remove singular regions, different masks were utilized in processing the data.

Similar displacement maps can be obtained for each level of load applied in the experiments. The diagrams in Fig. 5 show the distribution of displacements u_x and u_y that were experimentally measured along the control paths X_1-X_2 and Y_1-Y_2 illustrated in the figure.

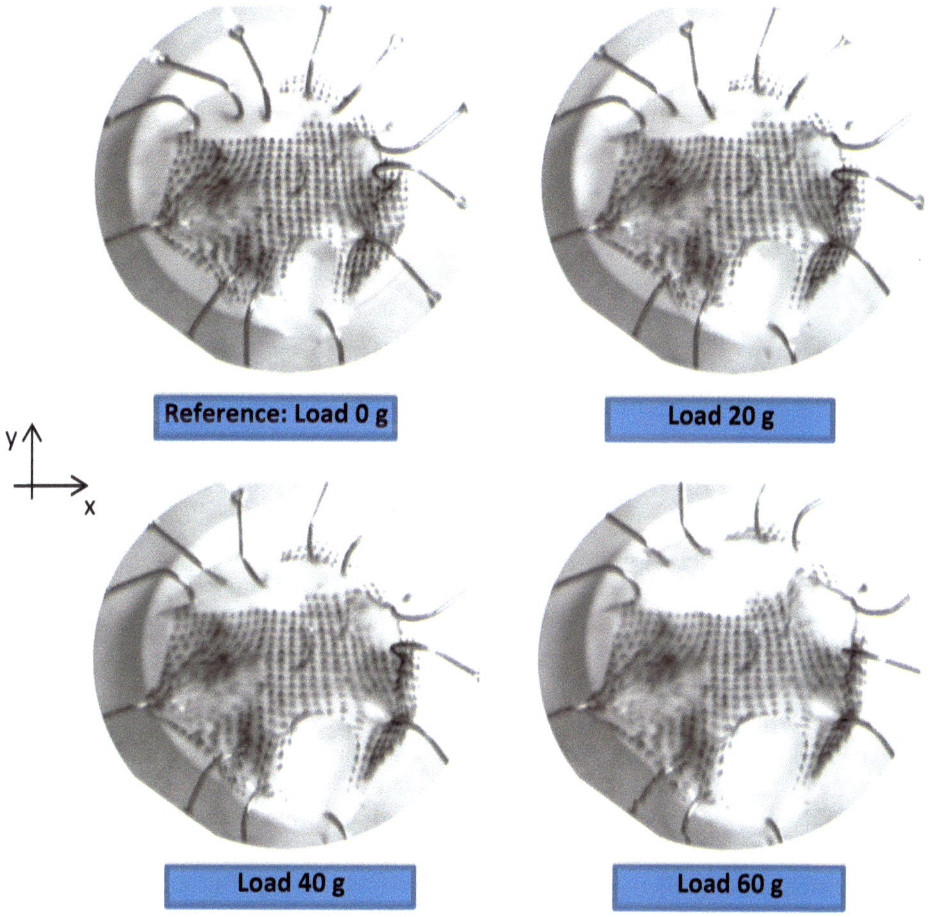

Fig. 2 Modulation of the dot grating printed on a slice of umbilical cord subjected to different load levels

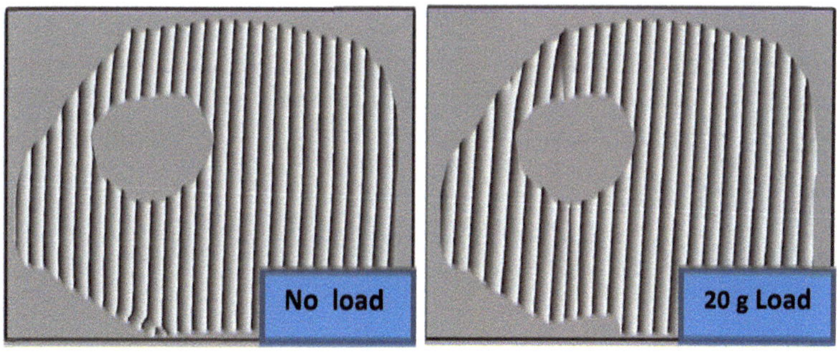

Fig. 3 Phase maps computed for the undeformed configuration and in presence of a 20 g load

Peri-placental samples resulted rather isotropic. Indeed, similar total *u*- and *v*-displacements (2.70 ± 0.19 mm and 3.59 ± 0.24 mm, respectively) yielded comparable values of the elastic moduli E_x and E_y (0.312 ± 0.085 MPa and 0.260 ± 0.069 MPa, respectively). On the opposite, peri-fetal samples resulted clearly anisotropic with different total *u*- and *v*-displacements (4.05 ± 0.32 mm and 1.26 ± 0.09 mm, respectively) producing the dissimilar elastic moduli E_x and E_y (0.383 ± 0.094 MPa and 1.267 ± 0.309 MPa). Overall, peri-fetal cord samples appeared significantly stiff along the y-axis.

Material properties were then fitted into hyperelastic constitutive models [5], respectively, 2-parameter Mooney-Rivlin for the Wharton's Jelly and third order Ogden for vein and arteries. An inverse analysis approach based on metaheuristic optimization was utilized for that purpose. Hyperelastic constants thus fitted were then given in input to the finite element model.

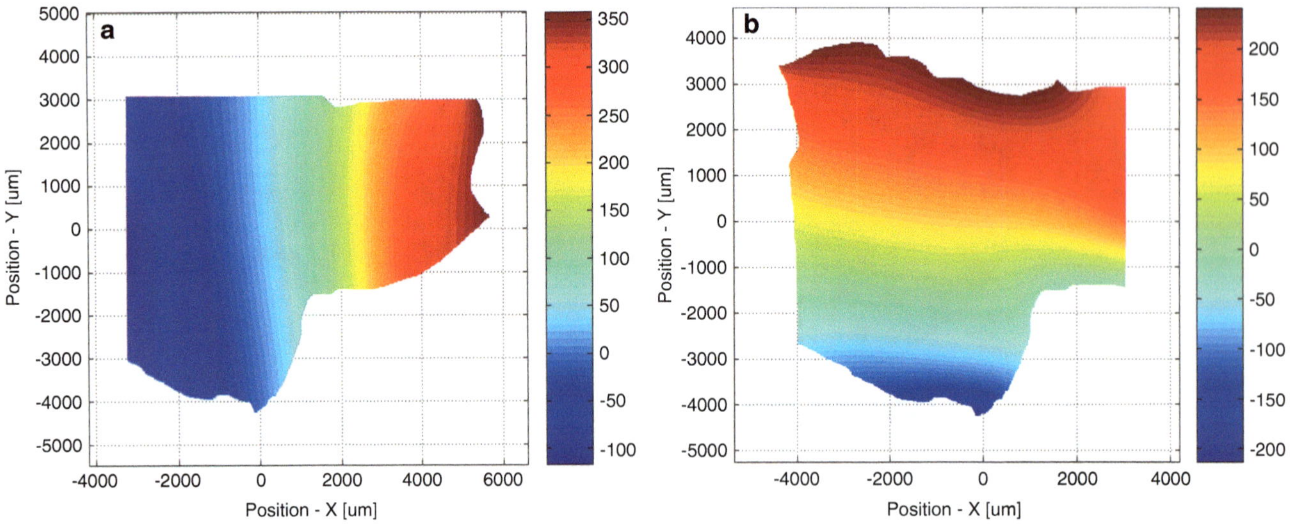

Fig. 4 In plane displacements (μm) computed along the X (**a**) and Y (**b**) directions in the sample for a load of 20 g

Finite Element Modelling and Analysis

FEM analysis of the WJ and vessel walls was performed in different simulated conditions of vessel endoluminal pressure and WJ elasticity. The maximum VM stresses developed in the walls of simulated straight vessels, when in presence of pressure regimens of 50 mmHg (arteries) and 20 mmHg (vein), were computed.

Average experimental values of material properties fitted from experiments were the input for the commercial finite element software ANSYS® (Version 14.5) [6]. WJ was modelled as a hyperelastic material since the largest strain values measured from moiré patterns were smaller than yield and failure limits reported in the literature [1]. In finite element computations, the large deformations option was selected due to the evidence that maximum relative displacements of diametrically opposed points approximated a maximum 20% of the entire specimen diameter.

Mesh refinement was achieved by convergence analysis (i.e. progressively reducing the number of model elements until the change of nodal displacements became smaller than 1%) and by using the SOLID92 10-node tetrahedral elements that include mid-nodes or vertices in the element sides to provide a quadratic approximation of the displacement field. Optimized 3-D finite element models of the WJ volume included an averaged number of 315,000 elements and 450,000 nodes. The 3D FE model along with a transverse section and the CAD view of the coiled arteries onto the vein are shown in Fig. 6.

Discussion and Conclusions

Intrinsic moiré was utilized to analyze the structural response of 2 mm thick slices of umbilical cord. In particular, the technique monitored the in-plane displacement field of slices subject to equibiaxial test. For that purpose, an ad hoc loading frame was designed and realized. Experimental tests showed that the mechanical behavior of the umbilical cord slices is highly anisotropic. The displacement field along the X direction is significantly different from that measured along the Y direction. The strong anisotropy of the tested samples can be justified with the anisotropic arrangement of the collagen fibers within the Wharton's Jelly as well as with the anisotropic disposal of the vessels within the sample.

Results of FE simulations indicate that the maximum arterial VM stress increases with raising values of arterial coiling and endoluminal pressure. Arterial and venous maximum VM stress did not appear to be directly correlated as the latter quantity is insensitive to arterial pressure. In the WJ, higher VM stress values always developed in the core region located between the arteries and the vein; however, the absolute values were up to 2.5 times greater in the peri-fetal portion of the cord. A reduction of arterial-vein distance as small as 1 mm profoundly altered the VM stress distribution, leading to a significant stiffening of the core WJ. Stress gradients disappeared as soon as arteries were modelled as straight vessels.

In conclusion, by using equibiaxial tension tests, intrinsic moiré technique and FEM analysis, we demonstrated a complex level of structural and functional hierarchy of the WJ and offered a direct visual appreciation of its pivotal roles in cord

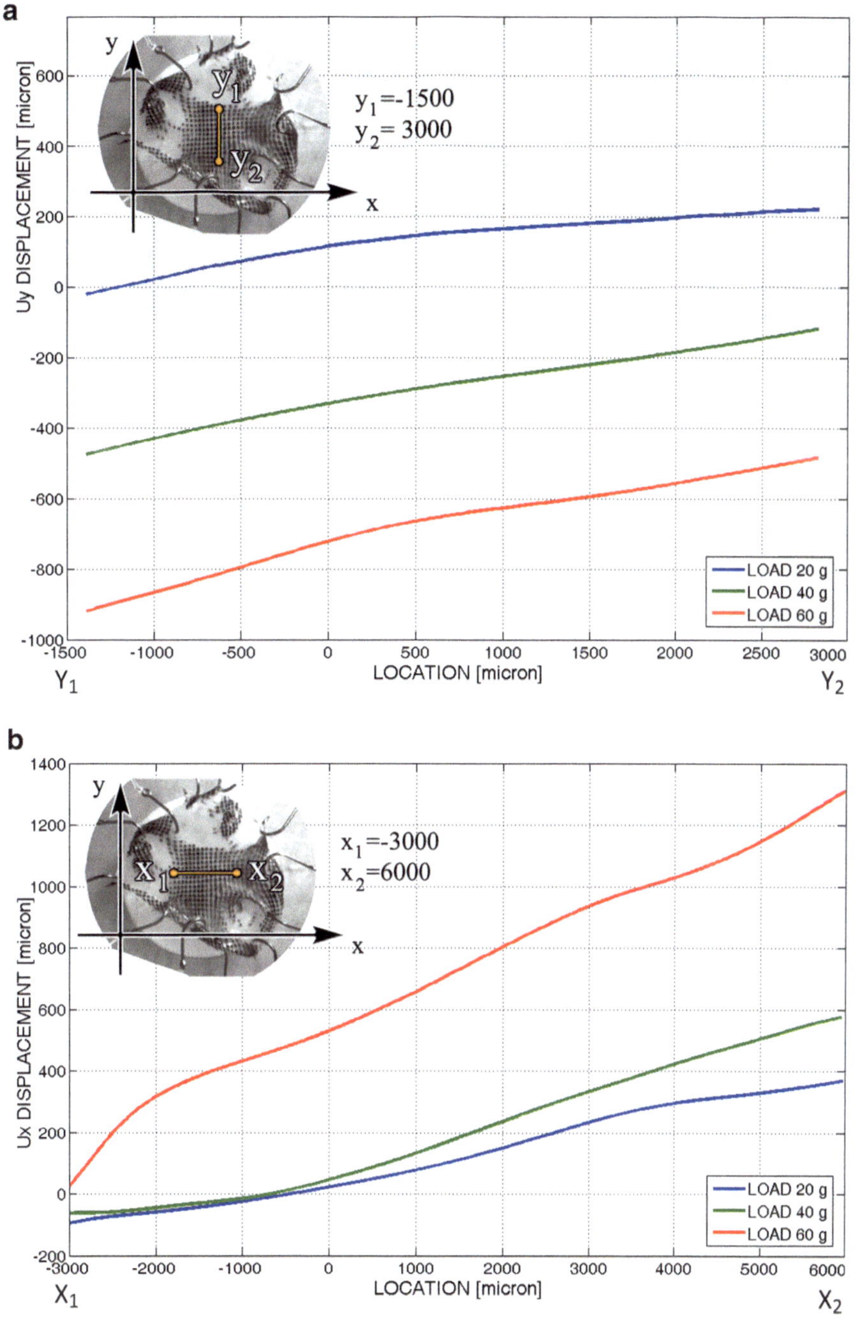

Fig. 5 In-plane displacements measured along the indicated paths in the Y (**a**) and X (**b**) directions for different values of applied load

hemodynamic and resistance to compression. Basically, WJ operates as a load transferring region based on its local stiffness: in the placental side, the softer WJ triggers the propelling effect of arterial wall pressure on the venous flow while the harder WJ retains stress in the perifetal side where propelling effect is no longer necessary.

Finite element models emerged as a suitable method to investigate the WJ and to obtain a direct understanding of the links between its biomechanics and pathophysiology in some high-risk clinical conditions such as fetal growth restriction and preeclampsia.

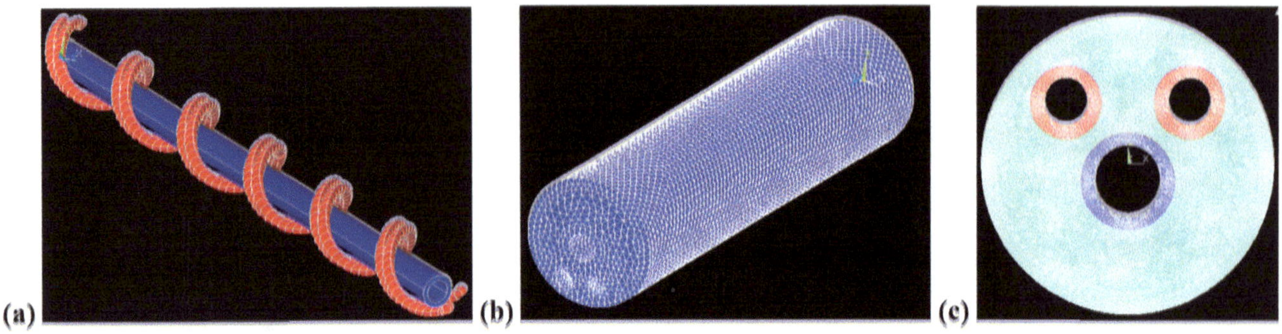

Fig. 6 (**a**) 3D model of spiralized arteries coiled onto the cylindrical straight vein; (**b**) 3D finite element model of the cord; (**c**) 2D finite element model of the transverse section of the umbilical cord

References

1. G. Pennati, Biomechanical properties of the human umbilical cord. Biorheology **38**, 355–366 (2001)
2. W.C. Li, X.Z. Ruan, H.M. Zhang, Y.J. Zeng, Biomechanical properties of different segments of human umbilical cord vein and its value for clinical application. J Biomed Mater Res B Appl Biomater **76**, 93–97 (2006)
3. B.A. Hamedani, M. Navidbakhsh, H.A. Tafti, Comparison between mechanical properties of human saphenous vein and umbilical vein. Biomed. Eng. Online **11**, 59 (2012)
4. R.W. Ogden, *Non-Linear Elastic Deformations* (Elsevier, New York, 1984)
5. C.A. Sciammarella, F.M. Sciammarella, *Experimental Mechanics of Solids* (Chichester, UK, 2012)
6. ANSYS® Version 14.5. *Theory and User's Manual*. ANSYS Inc., Canonsburg (USA), 2012

Laser Diffractometer for Measuring Bacterial Biodegradation of Dental Materials

Ying Gu, Philip Foo, Alan Guo, Austin Giordano, Stephen Walker, and Fu-Pen Chiang

Abstract *Streptococcus mutans* (*S. mutans*), a major cariogenic bacterium, has exhibited esterase activities, and has been shown to degrade dental resin composites and adhesives contributing to the risk of developing secondary caries and failed restorations. Changes to the surface morphology of the resin materials have been observed through SEMs; however, they were not quantified. In the current study, we present the development of a laser diffractometer whereby we can apply it to determining quantitatively the surface roughness spectrum of a dental material due to bacteria biodegradation.

Keywords Laser diffraction · Fourier optics · Biological systems · Surface metrology · Optical metrology

Introduction

Dental resin composites are popular restorative materials used in modern dentistry. This approach relies on the application of adhesive materials, which consists of a thin layer of resin applied between the conditioned tooth structure and resin matrix of the composite [1]. This adhesive layer effectively bonds resin composites to tooth structures to achieve a more conservative cavity preparation design. Although these restorations provide a more conservative and aesthetic treatment, their clinical longevity can be inadequate. The main cause of failure in composite restorations is the degradation of the adhesive interface over time, resulting in marginal leakage, which eventually leads to secondary caries, and subsequent loss of retention [1]. Therefore, the integrity of the adhesive bond layer is critical for the success of esthetic composite restorations.

Research has suggested that bacterial biodegradation can potentially contribute to the loss of adhesion and microleakage [2]. Esterase activities in human saliva can degrade bis-phenyl glycidyl dimethacrylate (BisGMA)-containing resin composites and adhesives. *Streptococcus mutans* (*S. mutans*), a major cariogenic bacteria, has such exhibited esterase activities, and has been shown to degrade dental resin composites and adhesives, increasing the risk of developing secondary caries and failed restorations. Changes to the surface morphology of the resin materials have been observed through SEMs; however, they have not been quantified [3]. In this current study, we present the development of a laser diffractometer whereby we can quantify the surface roughness spectrum of dental materials caused by bacterial biodegradation.

Y. Gu
Department of General Dentistry, School of Dental Medicine, State University of New York at Stony Brook, Stony Brook, NY, USA
e-mail: ying.gu@stonybrookmedicine.edu

P. Foo · A. Giordano (✉) · F.-P. Chiang
Department of Mechanical Engineering, College of Engineering and Applied Sciences, State University of New York at Stony Brook, Stony Brook, NY, USA
e-mail: philip.foo@stonybrook.edu; austin.giordano@stonybrook.edu; fu-pen.chiang@stonybrook.edu

A. Guo
Department of Civil Engineering, College of Engineering and Applied Sciences, State University of New York at Stony Brook, Stony Brook, NY, USA
e-mail: alan.guo.2@stonybrook.edu

S. Walker
Department of Oral Biology and Pathology, School of Dental Medicine, State University of New York at Stony Brook, Stony Brook, NY, USA
e-mail: Stephen.Walker@stonybrookmedicine.edu

M. E. Grady (ed.), *Mechanics of Biological Systems and Materials & Micro-and Nanomechanics, Volume 4*,
Conference Proceedings of the Society for Experimental Mechanics Series, https://doi.org/10.1007/978-3-030-30013-5_15

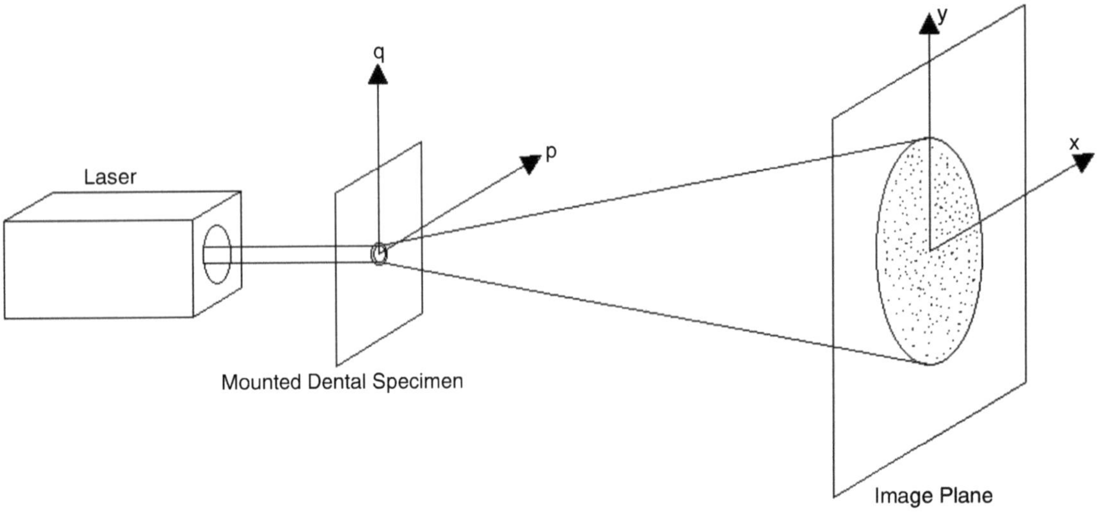

Fig. 1 Schematic of the Laser Diffractometer

Experimental Procedures

The design of the laser diffractometer is based on the principle of Fourier optics [4]. A schematic of the major elements of the laser diffractometer is as shown in Fig. 1. A lower power (10 mW) HeNe narrow laser beam with a wavelength of 632.8 nm and a beam diameter of 0.68 mm is used to illuminate a dental specimen which is mounted on a stand. A screen is erected at a sufficiently far away distance to receive the full diffraction pattern resulting from the surface roughness of the dental specimen. A camera is then used to record the diffraction pattern after ensuring that there is minimal light pollution for subsequent analysis.

It can be shown that when a coherent laser beam is impinging upon an optically rough surface, its far-field diffraction pattern is nothing but the 2D Fourier transform of the spatial frequencies of the surface roughness as described by the following Eq. (1).

$$F(p, q) = \iint_{-\infty}^{\infty} \mathrm{f}(\mathrm{x}, \mathrm{y})\, \mathrm{e}^{\mathrm{i}(\mathrm{px}+\mathrm{qy})} \mathrm{dxdy} \quad (1)$$

In which (p,q) and (x,y) are the coordinates at the specimen plane and image plane, respectively. Thus, by analyzing the diffraction spectrum of the laser beam one can deduce the surface roughness quantitatively. This characteristic has been used to assess the surface roughness created by plastic deformation of metals [5, 6]. In this study, we have applied this technique successfully to assess the surface roughness of dental materials resulting from bacterial biodegradation. The adhesive resin material we used in this study is Scotchbond Multi-Purpose Adhesive by the 3 M Company (Maplewood, MN, USA). Dental specimens were prepared with metal molds and Mylar strips, which were then photopolymerized for 20 s, 10 s on each side. Two experimental groups were established, with three specimens in each. The dental specimens were incubated in the presence (bacteria only group) or absence (control group) of bacteria *Streptococcus mutans* (*S. mutans*) for 4 weeks. Six specimens were then analyzed using the laser diffractometer technique and the resultant data compared. A digital microscope was used to record the visual profile of the surface of the specimens as well for comparison purposes.

Calibration Procedure

In order to quantify the spatial frequency spectrum of a rough surface, we use a diffraction grating of a known spatial frequency as a calibration gauge. By placing the calibration grating at the plane where the dental specimen is to be placed and illuminated with the same laser beam, the diffraction spectrum will be displayed on the screen and recorded by the camera. The calibration grating used in the study is a cross-grating of 1000 lines per in. in two perpendicular directions, or alternatively a diffraction grating of 1000 square dots per in. in two orthogonal directions (Fig. 2). The diffraction spectrum

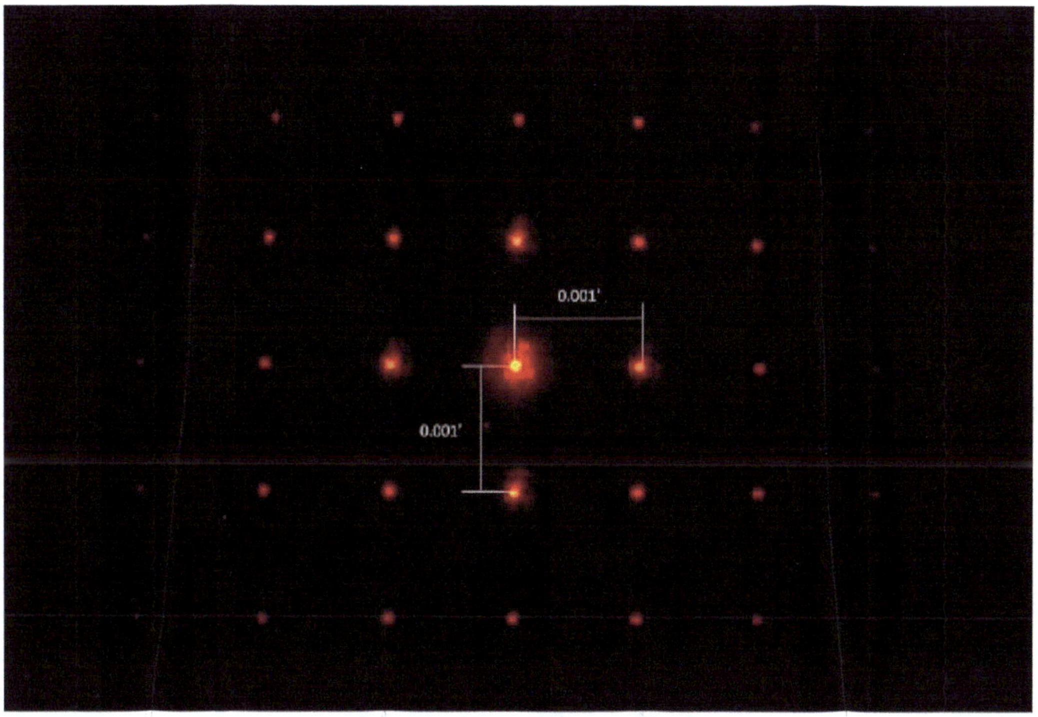

Fig. 2 Laser diffraction pattern from a 1000 lines per in. cross grating used for calibration

used for calibration is shown in Fig. 3. The distance between different diffraction orders represents quantitatively the spatial frequency of the grating (or the surface roughness of the dental specimen). As illustrated by the two calibration lines marked on the figure, each dot is at a distance of 0.001 in from any adjacent dot. Thus, by comparing the diffraction spectrum of a dental specimen subjected to a certain degree of biodegradation with that of the cross-grating, the surface roughness of the specimen can be quantified.

The central dot is indicative of the zeroth diffraction order, i.e. (0,0). This implies that any light which appears within that area originates from regions of the specimen that is flat and smooth. The first set of bright dots around the center is of the first order, and as such the first to the right of the central dot is of the (1,0) order and the first to the left is of the (−1,0) order. These two orders represent the fundamental spatial frequency of the diffraction spectrum of the grating lines along the horizontal axis. Similar interpretations can be made for the first two orders along the vertical axis, i.e. (0,1) and (0,−1), and the same applies for subsequent orders. Other smaller bright spots are of higher spatial frequencies of the diffraction grating and are of no concern in this study. In theory, if the spatial frequency spectrum of a grating is isotropic, the diffraction halo will manifest in the shape of a circle. Otherwise, it will in the shape of an ellipse or other shapes depending on the spatial frequency content of the surface roughness.

Results

In order to demonstrate the quantitative nature of the proposed technique, we applied it to mapping the roughness spectrum of dental specimens whose surfaces were biodegraded by the bacteria *Streptococcus mutans* (*S. mutans*) for 4 weeks. The specimens are in the shape of a circular disk of 0.0787 in. thick and 0.1968 in. in diameter. A thin Plexiglas plate and custom mount are used as a stand into which the dental specimen is secured. A digital camera is used to capture the diffraction pattern for subsequent analysis. As a comparison, a picture of the diffraction pattern from a virgin dental specimen from the control group (Specimen A) without bacterial degradation was taken first.

The result is shown in Fig. 3a. The shape of the diffraction spectrum is essentially in the shape of a circle, which means that the surface roughness of the virgin dental material is isotropic consisting largely circular cavities of various sizes. The minor axis of the diffraction halo is about 0.00392 in., or 3.92 spaces relative to the calibration diffraction pattern. It can then

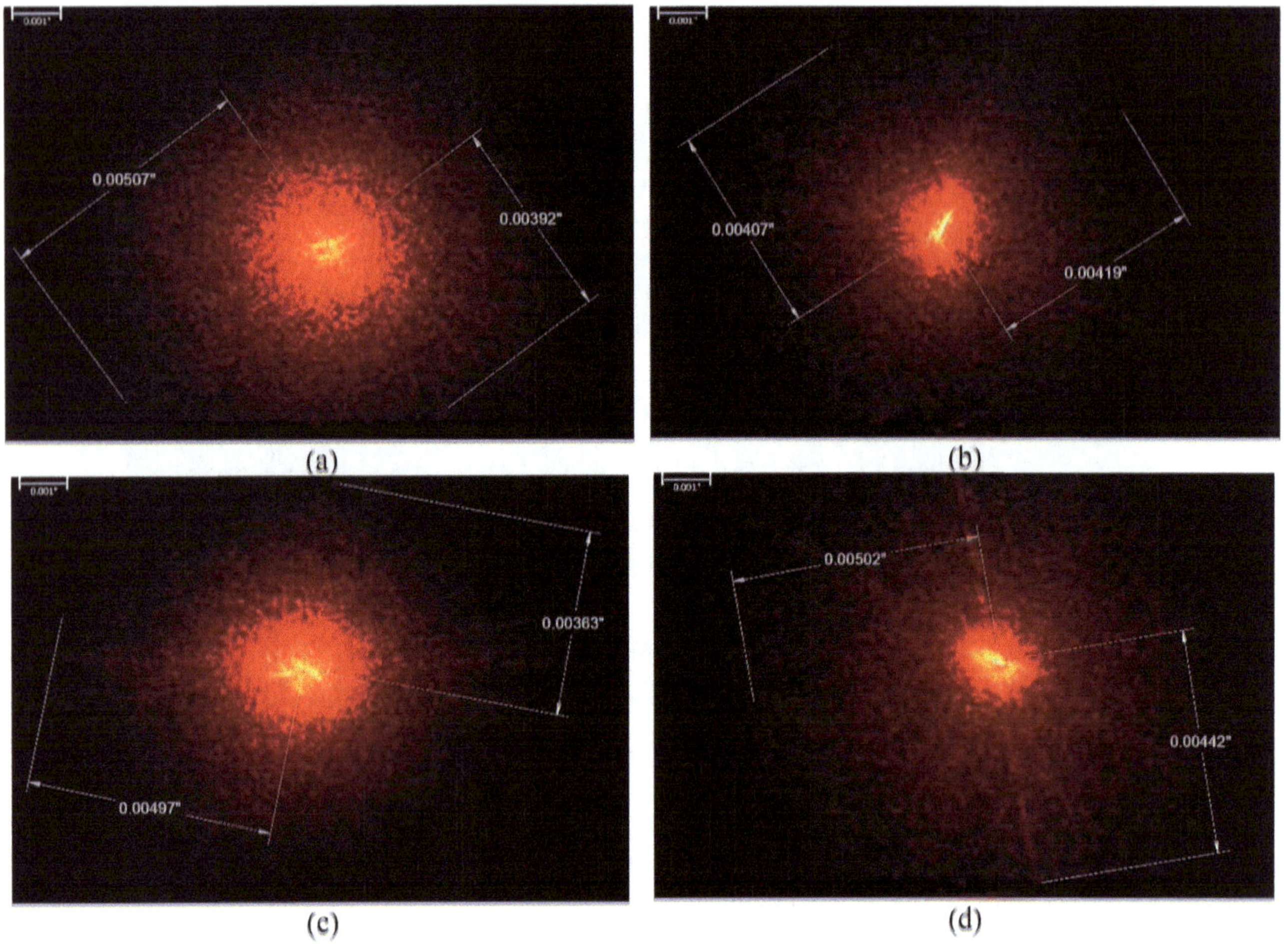

Fig. 3 Diffraction pattern of the specimen before and after incubation (**a**) Specimen A prior to incubation (**b**) Specimen A after incubation (**c**) Specimen B prior to incubation (**d**) Specimen B after incubation

be inferred that the smallest cavity is 0.001/3.52 = 0.000255 in. Following incubation, the diffraction pattern has remained circular, which is to be expected as no biodegradation has occurred (Fig. 3b).

Figure 3c, d are the diffraction patterns of the dental specimens before and after incubation with *S. mutans*, which is denoted as specimen B for the bacteria group. Figure 3c is similar to Fig. 3a, however, the diffraction pattern for the specimen following incubation exhibited a noticeably more elliptical shape as well as a broader spread, which implied that the surface roughness has changed due to the formation of numerous additional cavities. The smallest elliptical cavities on specimen B have the size with the major and minor axes being 0.0002 in. and 0.00023 in., respectively.

Intensity plots of the speckle pattern were generated, using Matlab to create a 3D plot where the x and y-axes represented the pixel location of the diffraction pattern, and the z-axis represented the intensity of the light. As shown in Fig. 4, all three specimens prior to incubation exhibited a relatively Gaussian distribution in their speckle intensity. For Specimen A, following incubation, the intensity plot retained the majority of its Gaussian nature as expected. For Specimens B, the intensity plots showed that the distribution is no longer Gaussian as indicated by the surrounding. The plots also showed that the diffraction pattern is not as isotropic following incubation as indicated by the significant decrease in the intensity of the central region due to the higher number of cavities as well as a greater variation in their sizes.

Characteristic photos of each specimen after incubation were taken using an optical microscope. Figure 5a, c represented the surface of Specimen A, and Specimen B, respectively. The lighting and magnification were kept constant for the characteristic photos. Figure 5b, d is composite photos for Specimen A and Specimen B, respectively. They are comprised of nine photos of the surface for each specimen which were taken separately and then compiled manually due to the limitations of the microscope used, and as such the lighting and color have small variations. These composite photos provide a visual representation of the area of the specimen through which the laser beam passed, and clearly indicate the expected surface profiles for each specimen and agree with the data provided by the diffraction patterns and the 3D intensity plots. Specimen A

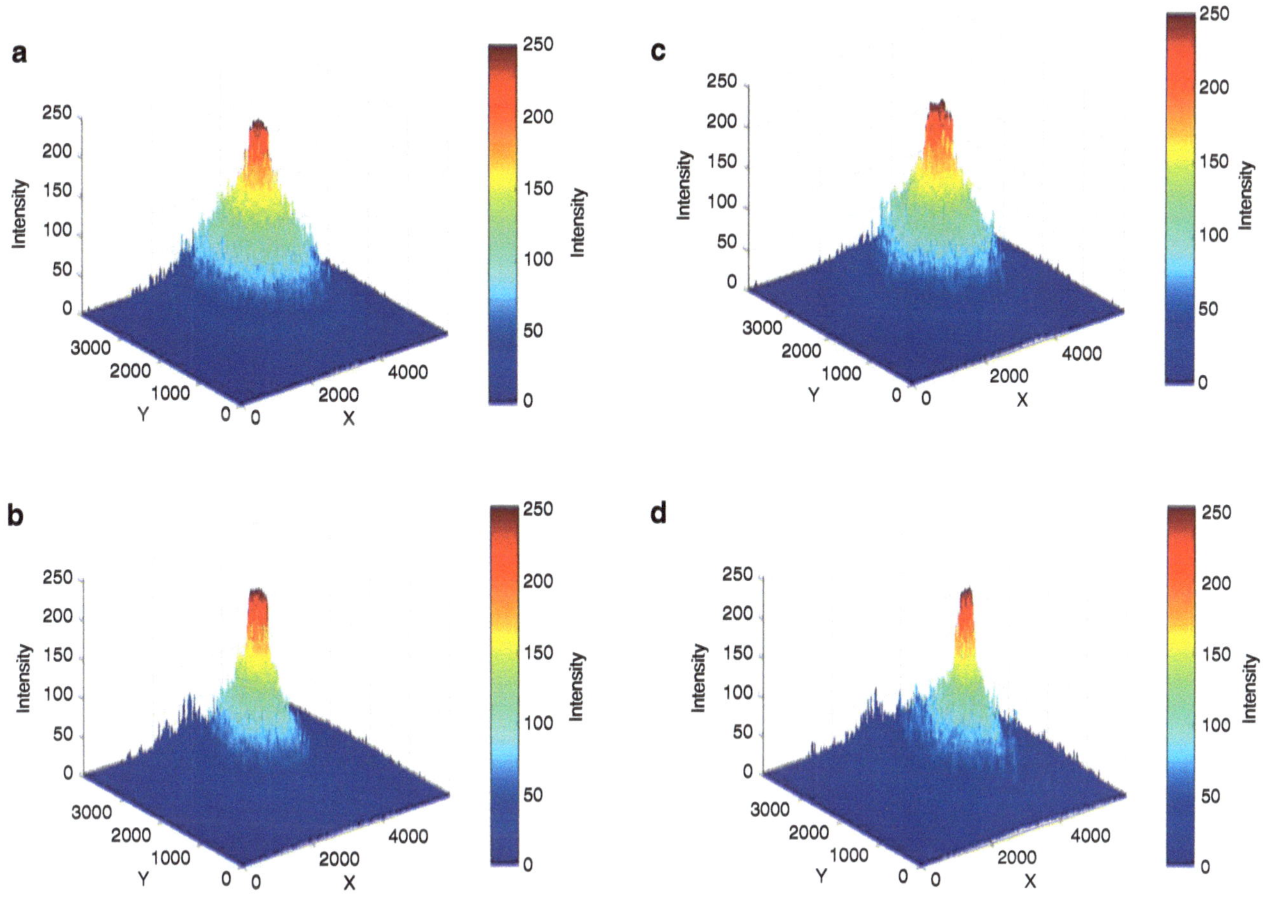

Fig. 4 Intensity pattern of the specimen before and after incubation (**a**) Specimen A prior to incubation (**b**) Specimen A after incubation (**c**) Intensity plot of Specimen B prior to incubation (**d**) Intensity plot of Specimen B after incubation

after incubation in the absence of bacteria has a relatively evenly distributed cavity size, while Specimens B after incubation with bacteria has a greater variation in cavity size and distribution.

Conclusion and Discussion

We have demonstrated that a simple laser diffraction technique can be used to delineate quantitatively the degree of biodegradation of dental materials. Variations were observed in the diffraction patterns obtained for all the specimens prior to incubation. This can largely be attributed to imperfections that developed during the manufacturing process of the specimens themselves. It should be noted that the Mylar strips used to cast the specimens were not optically smooth, and thus resulted in diffraction patterns that were not perfectly isotropic. These imperfections would also aid in facilitating the growth of bacteria colonies and would explain variations in the surface profile of the specimens exposed to the bacteria. Regardless, the expected changes in the surface roughness of the specimens due to varying levels of biodegradation were observed in the experimental group. Both the diffraction patterns and the 3D intensity plots indicate that the specimens which were subject to biodegradation only had a higher distribution of large-sized cavities and subsequently a rougher surface compared to those in the control group. The sharp aliasing that is observed in Fig. 3d is caused by scratches on the surface of the specimen, which can be observed in Fig. 5d, which can be attributed to manufacturing imperfections. The size and approximate shape of the cavities thus created manifest themselves in the size and shape of the laser diffraction pattern. By using a simple calibration procedure, the surface roughness spectrum can be easily quantified. The use of a scanning electron microscope to observe the surface cavities created by bacteria biodegradation is time-consuming and expensive. The laser diffractometer system proposed in this study is very simple and economical to set up. It is a very practical research tool to evaluate and quantitate the surface roughness spectrum of dental materials caused by bacterial biodegradation.

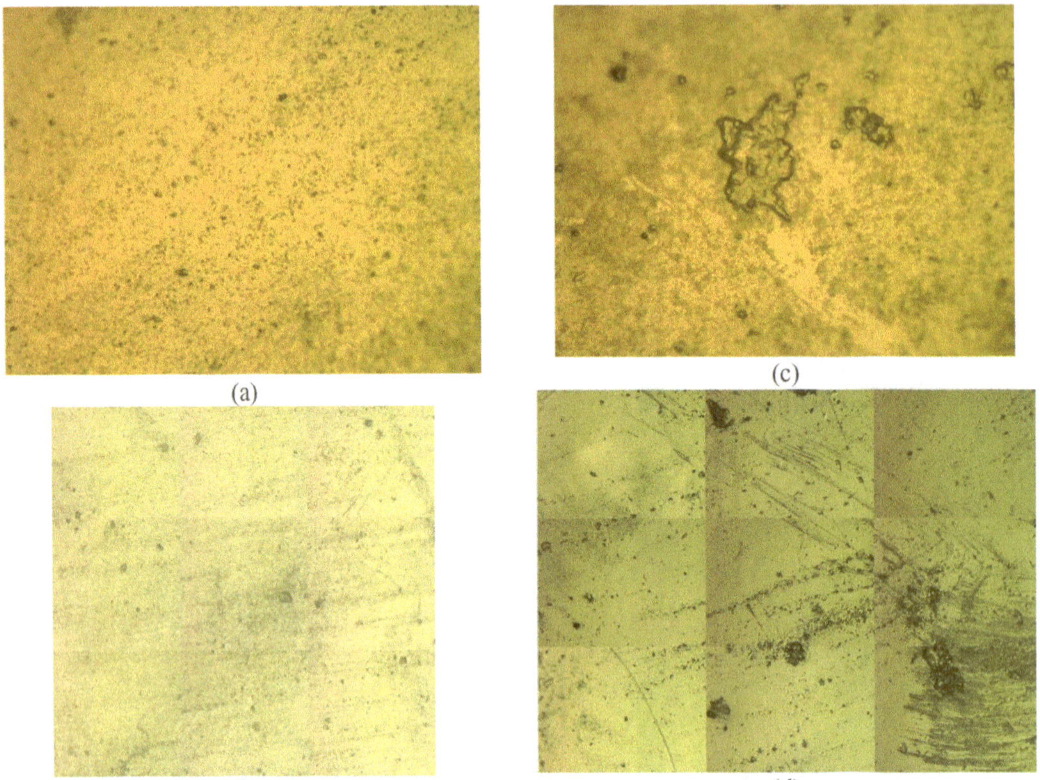

Fig. 5 Optical microscope photos of the specimen (**a**) Characteristic photo of Specimen A after incubation (**b**) **Surface** profile of Specimen A after incubation (**c**) Characteristic photo of Specimen B after incubation (**d**) Surface profile of Specimen B after incubation

References

1. Y. Liu, L. Tjäderhane, L. Breschi, A. Mazzoni, N. Li, J. Mao, et al., Limitations in bonding to dentin and experimental strategies to prevent bond degradation. J. Dent. Res. **90**, 953–968 (2011)
2. Y. Delaviz, Y. Finer, J.P. Santerre, Biodegradation of resin composites and adhesives by oral bacteria and saliva: A rationale for new material designs that consider the clinical environment and treatment challenges. Dent. Mater. **30**(1), 16–32 (2014)
3. M. Bourbia, D. Ma, D.G. Cvitkovitch, J.P. Santerre, Y. Finer, Cariogenic bacteria degrade dental resin composites and adhesives. J. Dent. Res. **92**(11), 989–994 (2013)
4. J.W. Goodman, *Introduction to Fourier Optics*, 3rd edn. (Roberts and Company, Englewood, CO, 2005)
5. F.P. Chiang, Y.Z. Dai, Assessment of flow stress and plastic strain by spectrum analysis. Exp. Mech. **31**(3), 197–201 (1991)
6. F.P. Chiang, Y.Z. Dai, Scattering from plastically roughened surface and its applications to strain assessment. Opt. Eng. **30**(9), 1269–1276 (1991)

MIX
Papier aus verantwortungsvollen Quellen
Paper from responsible sources
FSC® C105338

If you have any concerns about our products,
you can contact us on
ProductSafety@springernature.com

In case Publisher is established outside the EU,
the EU authorized representative is:
Springer Nature Customer Service Center GmbH
Europaplatz 3, 69115 Heidelberg, Germany

Printed by Libri Plureos GmbH
in Hamburg, Germany